TECHNOLOGY AND INTERNATIONAL STABILITY
(SWIIS 2003)

A Proceedings volume from the IFAC Workshop,
Waterford, Republic of Ireland, 3 – 5 July 2003

Edited by

P. KOPACEK
Vienna University of Technology,
Vienna, Austria

and

L. STAPLETON
Waterford Institute of Technology
Waterford, Republic of Ireland

Published for the

INTERNATIONAL FEDERATION OF AUTOMATIC CONTROL

by

ELSEVIER LTD

ELSEVIER Ltd
The Boulevard, Langford Lane
Kidlington, Oxford OX5 1GB, UK

Elsevier Internet Homepage
http://www.elsevier.com

Consult the Elsevier Homepage for full catalogue information on all books, journals and electronic products and services.

IFAC Publications Internet Homepage
http://www.elsevier.com/locate/ifac

Consult the IFAC Publications Homepage for full details on the preparation of IFAC meeting papers, published/forthcoming IFAC books, and information about the IFAC Journals and affiliated journals.

First edition 2004

Library of Congress Cataloging in Publication Data

A catalogue record for this book is available from the Library of Congress

British Library Cataloguing in Publication Data

A catalogue record for this book is available from the British Library

ISBN 978-0-080-44290-7

ISSN 1474-6670

Transferred to Digital Print 2010
Printed and bound in the United Kingdom

To Contact the Publisher

Elsevier welcomes enquiries concerning publishing proposals: books, journal special issues, conference proceedings, etc. All formats and media can be considered. Should you have a publishing proposal you wish to discuss, please contact, without obligation, the publisher responsible for Elsevier's industrial and control engineering publishing programme:

Christopher Greenwell
Publishing Editor
Elsevier Ltd
The Boulevard, Langford Lane   Phone: +44 1865 843230
Kidlington, Oxford   Fax: +44 1865 843920
OX5 1GB, UK   E.mail: c.greenwell@elsevier.com

General enquiries, including placing orders, should be directed to Elsevier's Regional Sales Offices – please access the Elsevier homepage for full contact details (homepage details at the top of this page).

IFAC WORKSHOP ON TECHNOLOGY AND INTERNATIONAL STABILITY 2003

Sponsored by
International Federation of Automatic Control (IFAC)
IFAC – TC 9.5. Supplemental Ways for Improving International Stability - SWIIS

Co-sponsored by
IFAC – TC 9.1. Economic and Business System
IFAC – TC 9.2. Social Impact of Automation
IFAC – TC 9.3. Developing Countries

International Federation for Information Processing (IFIP)
International Federation of Operational Research Societies (IFORS)

With the support of Coca Cola Bottlers,Ireland.

Organized by

Information Systems and Organisational Learning Research Group (ISOL)
www.wit.ie/isol

Waterford Institute of Technology

Institute for Handling Devices and Robotics (E318)
Vienna University of Technology

International Programme Committee (IPC)
Kopacek P. (AT) Chair

Brandt, D. (DE)
Cernetic, J. (SI)
Cioffi-Revilla, C. (US)
Dimirovski, G. (MK)
Dinibütün, A.T. (TR)
Dumitrache, I. (RO)
Groumpos, P.P. (GR)
Hersh, M. (UK)
Holubiec, S. (PL)

Kile F. (US)   US
Lu Y.Z. (CN)   CN
Makarenko, A (UA)
Mansour, M. (CH)
Scheffran, J. (DE)
Shubin, A (RU)
Stapleton, L. (IE)
Vamos, T. (HU)

National Organizing Committee (NOC)
Stapleton, L. (IE) Chair:

Brendan Lyng (IE)
Han M.W. (AT)
Kopacek P. (AT)
Nemetz I. (AT)

Patrick Wall (IE)
Sinead O'Leary (IE)

PREFACE

This anniversary conference marks twenty years since the SWIIS technical committee was established. It is held at a time of great trouble and uncertainty in the world, a time in which it is difficult for engineers and systems researchers to come together and share ideas, insights, dreams and vision. Barriers are being erected and finances are becoming more difficult to secure. People in many countries are afraid or unable to travel to meet their colleagues elsewhere in the world in events like SWIIS.

In these difficult times the work of SWIIS is more important than ever. It brings together researchers from many countries interested in understanding how technology can be used to benefit people across the globe. This conference presents a series of papers with a unique variety of perspectives and insights. The papers in this volume deal with issues as varied as globalisation, education, unemployment and economic systems. There are also two special sessions. One of these deals with the pressing need to revitalise the debate on engineering ethics, raising questions and providing new insights into how ethical issues can and must be addressed in a 'global village' and in the context of the geo-political structures of the twenty-first century. In the second special session the conference also brings together a series of contributions from the Russian Federation. It is now over ten years since the fall of communism in Russia and this session brings together unique perspectives from across the spectrum of current Russian intellectual activity in engineering and systems science.

The promise of technology so often remains elusive. In this twentieth anniversary of SWIIS the contributions to this set of proceedings volume shows that there are still so many opportunities for researchers as they attempt to harness technology for the benefit of all people of the world. The commitment to the original vision of the SWIIS committee remains as alive as ever.

L. Stapleton
NOC Chair and Editor

P. Kopacek
IPC Chair and Editor

CONTENTS

CASCADING UNDEREMPLOYMENT AND ITS EFFECTS

F. Kile

Microtrend
420 E. Sheffield Lane
Appleton, WI 54913-7181
USA

Abstract: Until recently, it was conventional wisdom that automation and machine-assisted thinking (AMAT) would provide more employment opportunities than it displaced. However, limits are emerging which are likely to end the surplus of new opportunities related to new technologies. Two factors underlie these limits: 1. Introduction into developing regions of technology beyond the capacity of these regions to adapt to its socioeconomic effects in an appropriate time frame; 2. Economic and environmental considerations constraining the growth in consumption that might otherwise accompany technological advances. As a result, AMAT may displace more jobs than it provides. These developments may trigger increasing underemployment and socioeconomic dislocations, raising a range of unprecedented ethical issues. *Copyright © 2003 IFAC*

Keywords: Automation, Economic, Environment, Global, Limits, Skill

1. INTRODUCTION

In 1811 a machine-smashing movement began in England. The machine-smashers were relatively skilled craftsmen. They were called Luddites after the legendary Nedd Ludd. They reasoned that their jobs would be replaced by machines. The Luddite movement was repressed. Also, the appearance of new employment opportunities due to industrialization weakened the rationale for destruction of machinery. Industrialization gave rise to new technologies and employment opportunities. Gradually, it became accepted wisdom that new technology would give rise to more jobs than it replaced. However, in the 21st century there is evidence that a new technology may no longer create more jobs than it replaces.

Three factors interact to suggest that technology no longer creates more jobs than it displaces.

a. Technology has slowly evolved from machines replacing muscle power to machines that replace a combination of human judgment and human physical activity. New technologies often demand higher levels of training than earlier technologies. The worldwide pool of people lacking adequate technical training is growing, while the pool of people with adequate training is growing less rapidly, if at all.

b. Employment in the developing world is heavily dependent on agriculture. Population growth drives people from rural areas into urban areas. Most of these rural people arrive without skills needed in high-tech industries in the cities. Unemployment and underemployment increase. As a result of these factors underemployment is beginning to grow more rapidly than unemployment.

c. A third, largely unrecognized factor may be the

most important driver of under- and un-employment. Burgeoning populations of urban poor in the developing world lack the means and opportunity to purchase the products of growing AMAT-driven industries. Most rural poor never participated in the increase in consumption which has been a growth engine for an increasingly productive global economy. Even if redistribution of income should create sufficient market demand for a flood of new consumer products, the global environment may not be able to support this level of consumption. There is ample evidence of environmental degradation caused by the present level of consumption.

If the reader agrees with our assumption that the global environment is unable to support present consumption without degradation, that same reader will conclude that a further increase in consumption will accelerate environmental decline. The ethical challenge associated with this assumption leads us to ask about the current generation's commitment to leaving a healthy environment for future generations. At some point (perhaps already reached) the absolute level of global consumption will either be capped or serious consequences will follow. The ethical imperative challenges us to cap aggregate loading on the global environment in the very near term.

This fact underscores the weakness of the Kyoto agreement on greenhouse gases. The Kyoto protocol provided for increases in total global output of carbon dioxide and other greenhouse gases. This implies that the Kyoto agreement was inadequate from the outset. Stronger agreements will be required -- along with real enforcement -- if the habitability of the planet is to be strengthened. Without strong action it is likely that aggregate human health will decline, with an increase in both morbidity and mortality.

AIDS in large areas of Africa has depleted an already inadequate pool of technically trained workers. Higher overall mortality and morbidity will not be concentrated in the same age group as AIDS, but any increase in deaths among the younger population will reduce the numbers of skilled people in the labor pool because younger people typically have more technological skills than the general population. Younger people are also more adept at learning new skills.

2. WHAT CAN/SHOULD BE DONE TO COUNTERACT THE ILL EFFECTS OF AMAT?

The problems of limits and displacements due to AMAT is new with respect to the world as a whole, though similar problems have long confronted populations in limited environments. Example include Eskimos in the inelastic Arctic environment and island populations. As limits affected these populations in the past, radical or novel solutions were often used. Alternatives to novel/radical solutions were starvation and mass deaths. Two examples illustrate the point:

a. In circumstances of environmental inadequacy, Eskimos long practiced a form of "voluntary suicide." This unusual code of ethics derived from the limited environment in which Eskimos live. Elderly Eskimos wandered off onto the Arctic ice and removed the burden of their care from their clans. They that they would soon die. In turn, they were aware that if they did not leave the clan, the entire community might die. A similar ethic prevailed among natives in the Tahitian Islands before the arrival of Captain Cook.

b. In the 19th century Ireland suffered from famine when the potato crop failed. The famine came while North America was accepting millions of immigrants, largely from Europe. Irish people migrated to North America in large numbers, reducing the death toll from famine. Mass emigration from a starving land to a friendly and hospitable region is no longer a serious option for large numbers. Moreover, the entire phenomenon of European immigration to North America was accompanied by unrecognized ethical issues. The ethical implications of displacing Native Americans from the only homes they ever knew were addressed much later, and only after the native population suffered mass deaths, humiliation and degradation. It is clear that any future mass migration will come at huge costs to both migrating populations and displaced populations. There is no remaining ethical justification for future large scale migration.
These examples illustrate that unusual measures must be taken to counteract environmental limits in a fixed domain.

3. THE WORLD AS A FIXED DOMAIN

Until recently, the world seemed more like an infinite domain than a fixed domain. Crowding, resource depletion, environmental loading and diminution of usable water supplies remind us that the world is a fixed domain while population and consumption behaviors continue as if occurring in an infinite domain. This suggests that, though individual decisions to consume various goods or to bear large numbers of children may appear ethical, the aggregate behavior of humanity in over-consuming and over-populating the earth is unethical, perhaps even to the extreme. Individuals, family units and nations are sub-optimizing their behaviors without regard to the larger whole. How can this mismatch of behaviors and environment be addressed? At the outset, the most acceptable approach is through education. An educational approach to matching human behavior with global capacity may be too gradual to be effective, but education can prepare people for a likely collision between aggregate human behavior and global carrying capacity.

The author noted in earlier papers that the environment does not "negotiate," irrespective of what international forums decide. The environment simply responds to stresses placed on it. To prevent accelerating environmental degradation and resultant effects on human well-being, aggregate global loading due to human activity must necessarily be capped.

Absent caps on environmental loading, future environmental consequences cannot be predicted. The need to cap environmental loading applies to all forms of loading, not merely greenhouse gases.

4. CAPPING CONSUMPTION

Moves to cap aggregate human loading on the environment will encounter feelings of entitlement in two major areas:

a. People in developed areas may argue that we live in a global economy and that they are entitled to whatever consumption they can afford. Moreover, they may argue that increased consumption increases employment. Arguments of this type beg the question regarding aggregate global environmental loading. If loading must be capped, as this paper argues, then consumption must also be capped. Some types of "consumption" do not increase environmental loading; example: if a TV screen is viewed by five people instead of by four, aggregate environmental loading is not affected. Clearly, "consumption" of this type is environmentally inconsequential. b. People may also argue that there is a basic right to procreate as much as one can. This is analogous to the specious notion that people are entitled to consume whatever they can afford. If loading on the global environment is to be capped, population must also be capped or per capita consumption must decline. In a milieu of increasing mass communication, psychological pressure to increase consumption is unavoidable. It may follow that only by *decreasing* total population can stable aggregate consumption be attained.

5. ALTERING HUMAN BEHAVIOR TO CAP ENVIRONMENTAL LOADING

Clearly, if aggregate environmental loading must be capped, individual consumption and total population must at some point necessarily be capped. This imperative is neither optional nor autocratic -- unless environmental distress is considered "autocratic" The requirement for caps reflects the finite and fixed environment in which we live. Our view of "rights" will change in the face of the unalterable parameters of life. We may feel we have a "right" to be exempt from laws of physics (such as gravity), but

"entitlement" to that "right" is illusory. The same is true of the so-called "right" to consume or to procreate. These "rights" are illusory when applied to the whole of humanity. Humanity must address the legal, ethical, and moral questions regarding how and by whom these "rights" are allocated among people.

This paper sets forth the hypothesis **that** some behaviors must change. It does not provide a framework for determining **how** behavioral changes can be achieved.

6. RECOGNIZING AND CONFRONTING UNDEREMPLOYMENT AND ITS ROOTS

What is underemployment? Underemployment occurs when people work below their skill level or with unduly low pay or work less than a desirable number of hours per week. This may be 36 hours or 40 hours or 45 hours. Outside of these bounds most people would feel that a person is either underemployed or overworked.

Underemployment is an added social burden beyond the demoralizing burden of being completely unemployed. Frequently, underemployment begins with unemployment. Persons who are completely unemployed because the work for which they were trained has been displaced by AMAT may accept work below a level at which their skills are used meaningfully. This of human skills may lead to loss of self-esteem as well as gradual loss of hard-won skills. Data regarding underemployment are scanty. Table 589 in the 2001 Statistical Abstract of the United States has a single entry for all part-time workers who are normally employed full-time. The entry is listed as "slack work or business conditions." It is common knowledge that far more people are employed below their skill levels, suggesting that little serious effort has been made to quantify data reflecting the seriousness of this issue. During travel to the Former Republic of Yugoslavia, I spoke with highly educated citizens of one of the successor states. They agreed that 50% of eligible workers are unemployed. There was evidence that some of the 50% unemployed were professional people working in a shadow economy, generally below a skill-level suitable for their level of education and training. For further discussion of this issue and comments on data which are available see the section Discussion of Relevant Studies below.

If a person displaced by AMAT accepts work below his/her skill level, that person displaces a less skilled person. The newly displaced person may accept a position beneath his/her skill level. Thus, underemployment cascades downward through a chain of skill levels at large costs to many people, each now working at a lower level than previously. In this way, an economy, which has achieved high productivity through applications of AMAT, may create both unemployment and underemployment across several sectors of the economy. As noted earlier, unused productive capacity, in terms of equipment and in terms of underutilized human capacity grows when potential customers have no money to purchase products or when they have inadequate opportunity to obtain or use these products. Another factor contributing to a likely increase in unused capacity is that the global environment has already exceeded acceptable limits on consumption, and that cumulative environmental stress will eventually cause a decrease in consumption. If other factors do not intervene first, rising death rates due to poor environmental conditions will reduce consumption.

If people respond voluntarily to environmental limits by decreasing consumption, underuse of productive capacity will increase. As production falls below a level equivalent to optimal economic activity, the cycle of unemployment and underemployment will be amplified.

7. ETHICAL DILEMMAS ASSOCIATED WITHOUR ACTION OR INACTION

There is no simple prescription for resolving the ethical dilemma accompanying decisions arising from uses of AMAT within a particular socioeconomic context. The accompanying web of related decisions is too complex even to develop guidelines for action or inaction. In any reasonable decision process, decision-makers, both individuals and groups, balance a range of options. A broad range of choices does not lend itself to precise, single-answer analysis. Instead, the decision-maker often sub-optimizes his/her problem solution. What is sub-optimization? Sub-optimization is analogous to the mathematical operation of partial differentiation. A sub-optimal decision maximizes one or a small set of benefits on a cost-benefit basis. A decision of this type does not necessarily involve a measurable quantity such as money. The decision may involve a tradeoff between a product manufactured to exacting specifications using AMAT but employing only a small number of people and a product manufactured to less rigorous specifications but employing many people. The ethical tradeoffs in a decision of this type frequently do not yield clear answers.

If we accept the premise that ultimately AMAT will displace greater numbers of people and that underemployment will cascade through a large labor pool, finally leading to virtually meaningless employment or total unemployment at the lowest skill levels, we may conclude that people need to be taught to live meaningful lives even when work is limited.

8. PROVIDING MEANING FOR UNEMPLOYED AND UNDEREMPLOYED PEOPLE

By current standards of employment (generally 36, 40, or 45 hours of work per week), more people in developed societies are overworked than at any time since rebuilding after World War 2 was largely completed. At the same time, in many nations, both developed and developing economies, unemployment and underemployment are almost out of control. Specific numbers are not cited for particular regions or countries, because the numbers are very fluid and because anything less than a complete list would likely offend some regions or their representatives. Anyone who reads large circulation newspapers or news magazine is aware of the depth and global scope of this double problem.

The following discussion of consequences of inaction and possible actions to reduce future social dislocations is speculative and intended to evoke discussion of issues not yet adequately addressed.

As an example of serious consequences from unemployment and underemployment, consider the situation in Germany in the late 1920s. The nation was in near chaos and there was little hope in sight. The way was paved for a demagogue to offer solutions. Hitler took advantage of the social disruption and, following his ascent to power after receiving only 33% of the vote in 1933, Europe suffered major upheavals and war. Perhaps if issues of un- and under-employment had been analyzed much earlier, the course of events would have been far happier.

In light of increasing prospects for un- and under-employment, any constructive discussion of related issues will be welcomed by many. Some potential remedies:

a. Allowing the situation to drift; in this instance, short-term costs are low, but long-term costs could be catastrophic. Dissatisfaction may lead to mass movements and social rebellions along with rising crime.
b. Imposing tight social control of employment, with reduced work levels for all people. The theory behind this type of social control is based on the notion that the environment can tolerate only a fixed level of loading and this could be a relatively even-handed mode of apportioning workloads and rewards. At least two problems would accompany this approach: Personal freedom would be reduced, and social control could be exercised by a small group of people, with a possible drift toward dictatorship. Free time would increase and problems linked with idleness would likely emerge.

c. Developing activities leading to personal satisfaction apart from "work life." Reaching this idealistic goal may seem unachievable, but discussion of this type of approach could lead to types of social organization congruent with a future in which limits to physical consumption and/or reproduction may be unavoidable. A major step might be development of an environmental ethic which persuades people that this is a desirable course of life in a fixed domain.

Some study has been done in the theory of cooperative games. This is a promising field of study which could lead to development of new forms of cooperative social organization. Moreover, even if satisfactory resolution of uneven distribution of employment in a fixed domain is attained, continued development of means of production suggests that the issue will re-emerge as aggregate human work decreases. This suggests that societal change will continue beyond initial successes in distributing work and providing for meaningful lives.
Past successes in dealing with prosperous societies have been based largely on rising incomes and consumption with accompanying high environmental cost. Continuous increases in consumption at the cost of relentless environmental degradation are not viable in the long-run. Society is challenged to develop meaningful, continuous change leading to an improved physical and emotional quality of life without ever-rising environmental loading.

9. DISCUSSION OF RELEVANT STUDIES

Relatively few empirical studies of underemployment have been published, and what is available was largely done since 1990. The world wide web was searched using an outstanding and exceptionally fast search engine "alltheweb.com." Of the references found several were commercial in nature, i.e., offers to assist underemployed people find better positions. This search engine is less commercial than some of its better known alternatives.

Over half of the approximately 70 non-commercial sites examined focused on small local areas. A more serious flaw in most studies was a heavy focus on people working 0-39 hours per week. Studies focusing solely on reduced work weeks (under 40 hours) were not examined further. Of the remaining few, four studies comprised the primary empirical base. Those four studies agreed on one major finding: underemployment is substantially more widespread than unemployment.

Empirical evidence from several sources validates theses advanced in this paper. The best study was done in Taiwan (Tseng), using year 2000 data. The publication date was not given, but it was clearly after 2000. This study indicated an unemployment rate of 2.3% and an underemployment rate of 30%. The study defined three types of underemployment; low hours (time-related inadequacy); low income (income-related inadequacy); and employment

mismatch (mismatch relating to education or experience). Employment mismatch was the largest group in the Taiwan study.

The other three studies, which also differentiated among categories of underemployed people, were done in the U.S. A study in Kansas (Glass) indicated an unemployment rate of 4.0% and an underemployment rate of 6.3% A study done in Nebraska, the state directly north of Kansas, yielded different data, surprising since the two states have similar demographic and economic characteristics. The Nebraska study indicated an underemployment rate of 29%. The sharp contrast between Kansas data and Nebraska data suggests that non-comparable measurement methods were used. A fourth study was done in several rural counties in Minnesota (Lawrence), a state whose demographic and economic characteristics are similar to Kansas and Nebraska. The Minnesota study did not differentiate in detail among the three categories of underemployment. However, the Minnesota categories corresponded with the categories used in the Taiwan study. The Minnesota study indicated a 22.2% rate of underemployment. That three of the four studies indicated roughly comparable rates of underemployment suggests that the fourth study (Kansas) employed different measurement criteria.

In conclusion, recent studies support the contention that underemployment is a serious labor issue. Analysis of these studies (especially the Taiwan study) suggests that the labor issue will continue to grow (a major thesis of this paper).

10. CONCLUSION: EMERGING ETHICAL IMPERATIVES

The suggestion that we might train people to live meaningful lives which are also unproductive by today's standards will seem like social heresy to anyone for whom a work ethic is important. Can we develop a "leisure ethic" which does not depend on increasing consumption patterns in an already burdened environment? The answer to this question is not apparent. Yet the likely increase of under- and un-employed people (UUP) poses serious ethical issues. If the UUP population grows, growing numbers of people could feel excluded from society. Enforced idleness has, in the past, fostered the growth of revolutionary or extremist tendencies or crime.

At the very least, a growing UUP group will marginalize people who are caught in a nexus of change. To avoid the growth of social radicalism and to provide meaning for people who do not "fit" into an ever-evolving society suggests that social experiments of many types will become an essential ingredient of a continually-changing social ethic.

REFERENCES

Glass, Robert W. (1996) The Effective Labor Force in Kansas: Employment, Unemployment, and Underemployment -- Executive Summary, Kansas, Inc. Topeka KS, USA.

Lawrence, Pareena G. Estimating Underemployment in West Central Minnesota, Stevens County Economic Improvement Commission, Stevens County, MN, USA.

Nebraska Dept. of Labor, Nebraska Underemployment Study, (2002) Lincoln, NE, USA.

Tseng, Shu-Fen, et al, New Economy, Underemployment and Inadequate Employment. Graduate School of Social Informatics, Yuan Ze University, Taiwan.

20 YEARS SWIIS
PAST – PRESENT - FUTURE

Peter Kopacek

Institute for Handling Devices and Robotics,
Vienna University of Technology
Favoritenstrasse 9 – 11, A – 1040 Vienna
e-mail: kopacek@ihrt.tuwien.ac.at

Abstract: The IFAC TC on "Supplemental Ways for Improving International Stability – SWIIS" is one of the longest situated in IFAC. According to first ideas during the IFAC World Congress in Kyoto 1981 this IFAC – at that time - Working Group organised in 1983 the first Workshop in this highly interdisziplinary field in Austria. Meanwhile a Technical Committee in IFAC, SWIIS was always a bridge between (control) engineers and various other disziplines to open IFAC to other related fields. Because of the 20th anniversary a short history of SWIIS as well as a outlook in the future, emphasising new topics, will be given in this paper. *Copyright © 2003 IFAC*

Keywords: Systems theory, systems engineering, conflict resolution, control of non - technical processes, history of IFAC.

1. BRIEF HISTORY

First ideas to install a IFAC Working Group on " Supplemental Ways for Improving International Stability" cames up in 1981 during the IFAC World Congress in Kyoto initiated by Hal Chestnut. As a result the first IFAC SWIIS Workshop was held in Laxenburg, Austria, Sept. 13 –15, 1983. It has greatly benefited from some of the international and interdisciplinary co-operations that were suggested in several of the presentations and discussions during the workshop. The major sessions of the workshop included the following topics: Cultural, Political, Educational, Behavioural, and Legal Aspects of International Stability; Techno-Economic Conditions for International Stability; System Analytical Approaches to International Stability; Negotiation and Mediation in Conflict Resolution; Decision-Making Processes.

The Second Workshop of the IFAC Working Group on SWIIS took place in Cleveland, Ohio, USA June 3-5, 1986. This workshop considered a most important question of our time - how can nations function without the need to go to war to settle international disputes?

In 1989 the IFAC/SWIIS workshop, International Conflict Resolution using Systems Engineering, was organised by the Computer and Automation Institute of the Hungarian Academy of Sciences in co-operation with the Austrian IFAC NMO. This theme thus continued a focus on the interrelationship between technology and conflict resolution that had been established at the previous workshop held in Cleveland USA in 1986.

One of the most important activity was the SWIIS' 92 Workshop. It was held in Bolton (Toronto/Canada), September 21-24, 1992 and was organised by "Science for Peace" on behalf of Canadian National Committee for IFAC. The program of SWIIS'92 contained over 20 technical papers from eight different countries.

The changes affecting the SWIIS Working Group that arose from the 1993 Congress in Sydney present a useful and welcome opportunity to structure and focus the SWIIS activities for the coming years.

While the indicated and probable elevation from WG to Technical Committee status is in fact mainly just part of a general IFAC restructuring that is taking place, the modified scope suggests some opportunities for new activities and possible changes in direction for SWIIS and its members.

The IFAC event on „Supplementary Ways for Improving International Stability" - SWIIS'95 was held in Vienna, Austria from September 29th to October 1st, 1995. This fifth event in the SWIIS series was organised by the Institute for Handling Devices and Robotics of Vienna University of Technology. Meanwhile the working group SWIIS was a Technical Committee (TC) in IFAC and the triennial workshops were appointed as regular conferences.

International stability refers to conditions in which nations, in an interdependent way, interact with one another in ways, which permit gradual changes with time in a mutually acceptable scale and direction. This development under stable conditions is considered with respect to social, political, ecological, national and international, regional and global aspects.

The conference continued the tradition set in the earlier four SWIIS meetings. The goal was the beneficial application of systems engineering methods onto description of conditions, in which nations or groups interact with one another. Scientists from other fields such as political science, economics, social science, and international studies should have a platform to present and discuss their ideas. Perhaps, this SWIIS event differed from earlier SWIIS meetings in the efforts to attract a younger generation to the work in the framework of this TC.

Organised by the Technical Committee, a SWIIS session entitled „Supplemental Ways for Improving International Stability" was scheduled by the organisers of the IFAC World Congress in San Francisco, 1996. All five presented papers gave an excellent survey of the scope of our committee.

According to the SWIIS TC meeting in San Francisco 1996 the 7th SWIIS conference was held in May 14-16, 1998 in Sinaia/ Romania. Papers were given in the following areas: ethodological analysis investigation of development: stability, sustainable development; modelling of stability; application of control principles to international Stability; East/West/North/South relationships; International policy co-ordination; global development: regional impact; cultural and political aspects in International Stability; educational and behavioural aspects; negotiation and mediation in conflict; applicability of the systems concept.

On the IFAC World Congress 1999 in Beijing the TC SWIIS was responsible for the organisation of two technical sessions. Both had with approximately 40 a very high attendance. The trend, inclusion of more economical and historical topics, started at the two last SWIIS events in Vienna and Sinaia was continued. Examples were the presentations from

Dimirovski and from one Institute of the University of Klagenfurt. Starkermann gave very exciting speeches on application of multivariable systems theory to conflict situations.

On the TC meeting it was decided to have an intermediate SWIIS workshop in Macedonia. One of the reasons was the actual political situation in this region. Therefore we had, at the first time in the SWIIS history, contributions of colleagues from Macedonia. This Workshop was very successful and on the TC meeting it was decided to co-operate closer with the the IFAC TC`s on " Social Effects of Automation" and " Developping countries". As a first result of this decision SWIIS was responsible for the organisation of two invited sessions on the DECOM event 2001 also in Macedonia.

The last IFAC Conference was organised in Vienna in 2001 attended by colleages from SOCEFF and DECOM as well as a large number of scientists from Russia. At the IFAC World Congress 2002 in Barcelona SWIIS was responsible for the organisation of a special session. On the TC meeting it was decided to have the next SWIIS conference in Waterford (Ireland) in 2003 and a "Multitrack Conference" in September 2004 in Vienna.

2. CONCLUSION AND FUTURE DEVELOPMENTS

As pointed out earlier one of the original ideas of SWIIS was to contribute with system theoretical and systems engineering methods to conflict solution. The SWIIS community started with the classical approaches of control engineering especially control of time continuos systems like theory of linear or some times non-linear systems, modelling, stability, optimisation. In the history of SWIIS there were some new approaches presented on several events for application of new methods from control engineering to SWIIS problems. Examples are multivariable and timevarying systems as well as fuzzy and neuro methods.

An other new approach to the SWIIS problems is the use of methods from manufacturing automation – time discrete systems – as well as the improvement of the interdisciplinarity. This is one of the tasks for the future.

3. LITERATURE

Preprints of the IFAC SWIIS Workshop "International Conflict Resolution using Systems Engineering", Budapest, 1989.

Kopacek P., H.Chestnut, J. Scrimgeour and F. Kile (1992): A Review of the Activities of the IFAC SWIIS Working Group 1983-1989. *In:*

Preprints of the SWIIS`92 Workshop, Toronto, Canada.

Kopacek, P.: SWIIS – An important expression of IFAC Commitment to Social Responsibility. *Proceedings of the IFAC Workshop "Instability Resolution in Regions of long confronted Nations – SWIIS 2000"*, Ohrid (MAC), May 2000, p.13 – 16.

ELSEVIER

IFAC
PUBLICATIONS
www.elsevier.com/locate/ifac

GOAL SETTING AND WORKING OUT OF THE STRATEGY OF DEVELOPMENT OF SOCIO-ECONOMIC OBJECTS

Z. Avdeeva, S. Kovriga, D. Makarenko, V. Maximov

*Z. Avdeeva - Institute of Control Sciences of Russian Academy of Sciences,
PhD Student;
S. Kovriga - Institute of Control Sciences of Russian Academy of Sciences,
PhD Student;
D. Makarenko - Institute of Control Sciences of Russian Academy of Sciences,
PhD Student;
V. Maximov - Institute of Control Sciences of Russian Academy of Sciences,
Head of Sector of Cognitive Analysis and Situation Modelling, PhD.*

Abstract: Present paper is dedicated to examination of goal setting and technique of structure and goal analysis and their application to working out of strategies of purposeful development of complex objects under instability. *Copyright © 2003 IFAC*

Keywords: Goal setting, structure and goal analysis, cognitive science, cognitive technologies, economic systems, forecasts, modelling

In conditions of rapid technological changes the requirement to quality of control of purposeful development of complex social and economic objects (SEO) (states, regions, transnational corporations etc.) dramatically increases at all levels.

One of the most important stages of control of a SEO is the goal setting that represents a process of modelling of result of yet not carried out activity, submitted more often in the image, mental model of the future condition of a situation, qualitative either quantitative characteristics or system of symbols and concepts.

At a stage of a goal setting it is rather important to reveal contradictions in goals that can be caused either by really contradictions or by chosen ways of reaching of goals.

The aim of goal setting is revealing of laws of goal formation and working out of the most effective ways of their reaching.

Goal setting includes:

- determination of vectors of goals of SEO development;

- analysis of opportunities of the set goals reaching;

- selection of means and ways of goals reaching.

1. GOAL SETTING WITHIN THE FRAMEWORK OF DECISION SUPPORT TECHNOLOGIES

Computer modelling of goal setting processes is becoming one of the basic directions of decision makers support. There has been developed a great amount of software explaining the past, but not capable to produce recommendations on solving of problems that can arise in the future.

This situation has resulted in necessity of creation of integrated systems of support of goal setting process and making of administrative decisions when working out a strategy of SEO development. Making decisions involves processes of great complexity. Experience testifies that decision support systems (DSS) that include a goal setting stage provide decision making for shorter period. It becomes possible because a decision maker has a model he can refer to when generating various alternative strategies of purposeful SEO development (goal setting models), and only then he considers and estimates the variants of decisions. This shift to the initial stage (generation of alternatives guiding by models) considerably changes the whole paradigm of decision making.

The modern process of goal setting consists of the following stages:

The first stage – analysis of environment from the point of view of the circumstances that have caused the necessity of decision making, i.e. in other words reconnaissance;

The second stage – search, working out, and analysis of possible variants of actions (alternatives), in other words planning;

The third stage - determination of a concrete sequence of actions out of possible ones, in other words selection.

The stages boundaries are indistinct, and the stages can be detailed.

When setting the goals of SEO development the decision makers should simultaneously be engaged in reconnaissance, planning, and decision selection.

Complexity of analysis and forecasting (modelling) of purposeful SEO development under instability of environment is caused by the fact that any action directed to reaching of the set goals, causes a number of consequences that can in turn prevent from goal reaching.

The number of factors describing such situations can be measured by tens and all of them are plaited in a web of cause and effect dependences. It is extremely difficult to see and realize the logic of development of events on a such multifactor field and at the same time it happens to change the goals and make decisions on the measures providing development of a situation in the necessary direction.

In such conditions it is more expedient to have an opportunity to make a qualitative concept of a situation as a whole at a level of tendencies of evolution of the processes describing a researched situation instead of searching for correctly formulated but rather laborious precise solutions of the quantitative

problems, concerning particular aspects of the situation under research. Such "qualitative" picture of a situation development as a whole is more useful when working out a forecast in order to substantiate the purposeful decisions, than attempts to construct exact, but particular model of happening changes. Thus, the shift to qualitative modelling, process structuring, and decision making on the basis of qualitative models is a reasonable alternative.

2. THE STRUCTURE AND GOAL ANALYSIS OF A SEO DEVELOPMENT

When setting the goals of a SEO development a decision maker doesn't always manage to trace if the goals he has set are inconsistent, i.e. reaching of a goal will prevent from reaching of another one. Inconsistency of goals can also be influenced by the chosen ways of their reaching.

It is caused by:

- First, that goal setting affects the interests of a situation participants that (interests) can not coincide with each other.

- Second, that it is rather difficult to take into account the structure of indirect influences between the factors of a situation that can be a source of latent contradictions in goals.

Thus, it is very important to reveal the contradictions already at the stage of goal setting.

The Institute Of Control Sciences of the Russian Academy of Sciences has developed a technique of the structure and goal analysis of a SEO development. The technique is a part of developed technology of cognitive analysis and modelling (Maximov V., 2001; Maximov V., Kornoushenko E., Makarenko D. 2001) and allows to determine integrated (direct and all possible indirect) influences of one factor on the other and due to it to reveal inconsistencies between goal and control factors. The structure and goal analysis also allows to determine the most effective controls, i.e. those from control factors which "are more effective" due to their integrated influence on the goal factors.

Thus, the structural analysis of cognitive model of a situation development under control consists of the following stages:

Stage 1 - analysis of goals (coordinates of a vector of goals) on mutual consistency in order to answer the question "whether the vector of goals (fixed or unfixed) is inconsistent, i.e. whether the reaching of any of goals (coordinates in a vector of the given goals) will prevent from reaching of other goals?".

Stage 2 - check of a consistency of the set of control factors with the given vector of goals, i.e. whether the change of the value of any control factor (with the help of the appropriate control) will promote reaching of some goals in a vector of goals and at the same time prevent from reaching of other goals of a vector of goals.

Stage 3 - estimation of efficiency of influence of control factors on all coordinates of the vector of goals. Such estimation is useful when choosing the most effective control factors the changes of which with the help of the selected control actions will provide the purposeful development of a situation.

The general technique of cognitive analysis and modelling and the structure and goal analysis is depicted on Fig. 1.

The structure and goal analysis enables the analyst:

- to remove the internal contradictions of strategies of a complex object development, i.e. to set consistent goals out of a set of goal factors and to select consistent controls out of the possible vector of control factors;

- to use the opportunities concealed in a situation structure in order to select the vector of control factors that is potentially effective in respect to goal reaching.

2.1. Analytical basics of the structure and goal analysis

The analytical basis of structural and goal analysis is made up by the cognitive approach (Maximov V., 2001) to structuring of knowledge of a SEO development. On the basis of such structuring one can create a cognitive map of a situation. The map determines the structure of interaction of factors of SEO internal environment with a boundary layer of environment.

Cognitive map represents a square table, in which:

- lines and columns correspond with the basic factors of a situation in terms of which the processes in a situation are described in a one-to-one manner;

- the element that is situated on crossing of line "i" and column "j" reflects the fact of direct influence of the factor "i" on the "j" one. Sign of this element displays a sign of influence (positive or negative), and the module - strength of such influence in the appropriate scale.

Cognitive map is the initial static representation (reflection) of connections between the factors existing in a situation under research. To solve the problem of a SEO purposeful development that arise in poorly

structured situations it is necessary to construct a dynamic simulation model and on its basis to obtain a new knowledge of the structure and dynamics of a situation under research.

Analysis of a graph model of a situation associated with a cognitive map allows to reveal the structural properties of a situation. The basis of the model is a weighed digraph $G = (X, A)$, where X is a set of nodes that biuniquely corresponds to the set of basic factors, A is a set of arcs reflecting the fact of direct influence of factors. Each arc connecting some factor X_i with some factor X_j has the weight a_{ij} which sign depicts the sign of influence of the factor X_i on the factor X_j, and the absolute value of a_{ij} depicts the strength of the influence. Thus the cognitive map can be examined as a connectivity matrix A_g of the graph $G = (X, A)$.

While the situation evolves each factor is being influenced not only by "neighboring" factors, but also by more "distant" ones and these indirect influences are transferred through chains of the appropriate factors and graph arcs that connect them. Set of influences as direct, and indirect to which each factor in a situation is subject is described with the use of concept of transitive closure of a cognitive map of the situation, determined as the sum of infinite power series

$$E_n + A + A^2 + ... + A^t + ... \quad (1)$$

of matrix A. Each element of this row characterizes passage of routes of length "t" in the graph, i.e. realization of direct and indirect interferences through one factor, two factors, etc.

Estimation of the sum of this series can be obtained only if the graph G adjacency matrix is stable. Then all elements of this series approach to finite limits at unlimited increase of t.

To determine the transitive closure it will suffice to consider N terms in a power series of matrix A, where N - the order of matrix A, i.e. number of basic factors in a cognitive map of a situation. Then the transitive closure of matrix A is estimated by matrix:

$$Q = E_N + A + A^2 + ... + A^N \cong (E_N - A)^{-1} \quad (2)$$

Therefore it is necessary to stabilize graph G of formal description of a situation (Maximov V., Kornoushenko E., 2001).

The goal of a situation development is described by a subset of goal factors of cognitive model. That means that the vector of goals of a situation development is a vector of values of goal factors (fixed goal), or a vector of directions of change of these values (unfixed goal).

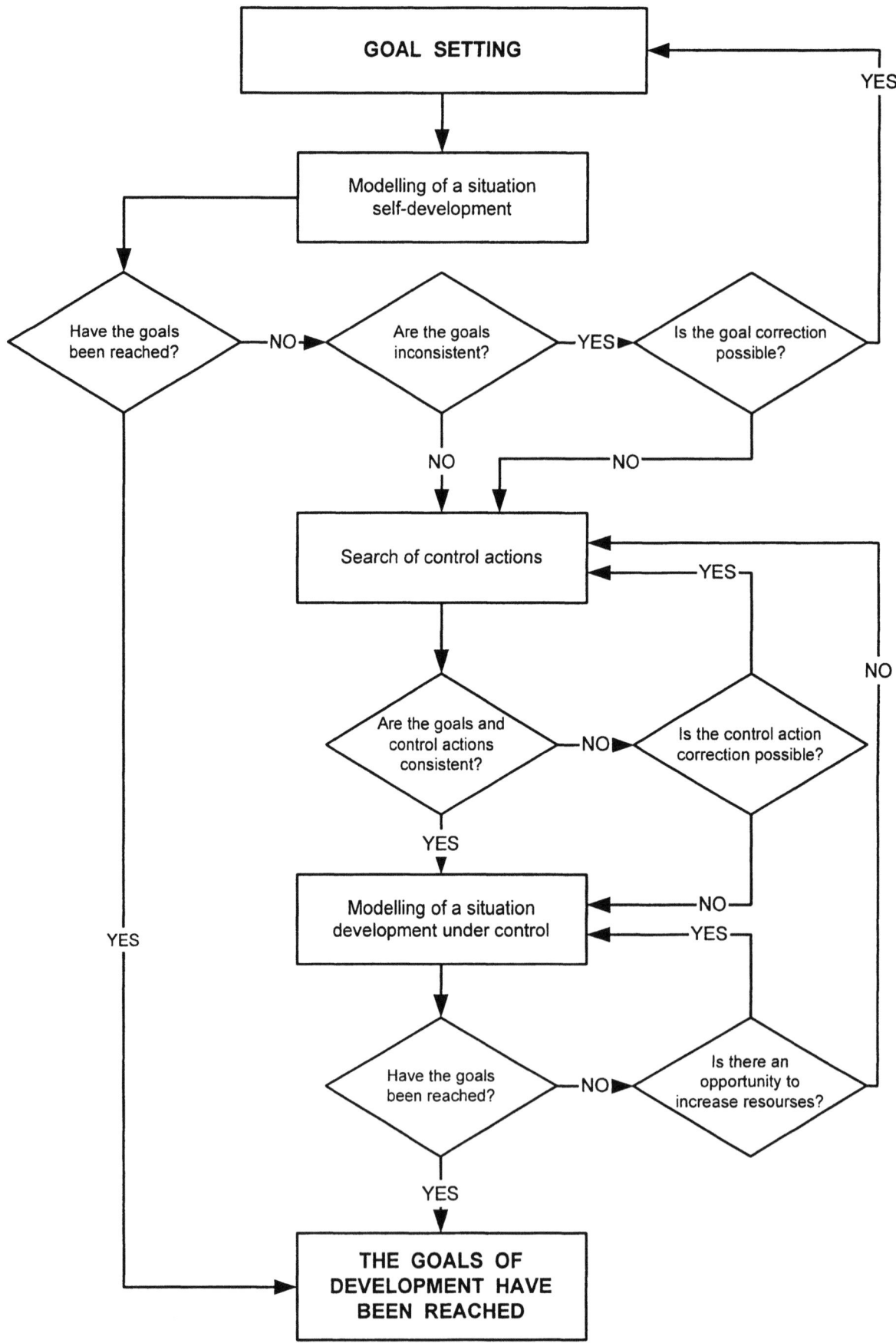

Fig. 1 The general technique of cognitive analysis and modelling

The fixed vector goal includes vector of goals Y^* and preset values of trends of change of each goal Y_i^*.

The fixed vector goal is a point in m-dimensional space of trends of goal change. In other words, goal is a vector of some "ideal" values of trends of goal factors change.

Goal setting hypothesis: a decision maker or an analyst can determine what direction of a factor change (the most significant factors for them) is desirable. The desirable direction of change of the factor x_i is determined by a parameter (estimated value) r_i, that possesses the value $+1$ if the increase of value of the factor x_i is desirable, and -1 if reduction of the factor x_i is desirable.

If the analyst finds difficulty in determination of a desirable direction of the factor x_i change, r_i is considered to be equal to 0. The parameter r_i refers to as the estimation of the factor x_i change (EFC).

Definition 2. The unfixed vector goal includes vector of goals Y^* and directions of favourable trends of change of its coordinates according to their EFCs.

Vector of favourable trends is a vector of interests of a decision maker (analyst). Restrictions are not imposed on the value of favourable change of goal factors (the more - the better).

Analysis of a goal vector coordinates on consistency. Let $Y = \{y_1 ... y_m\}$ be a set of goal factors and $r(Y)$ - a vector of desirable EFCs.

Definition 3. Vector of goals Y is consistent if

$$r_j r_k = sign(q_{ik}) \text{ for any } y_i, y_k \in Y. \qquad (3)$$

where q_{ik} - (i, k)-th element of matrix Q.
If (3) is fulfilled for goal factors y_i, y_k they refer to as consistent, otherwise these factors are inconsistent.

When the consistent vector of goals is formed the desirable integrated change of any of goal factors will not result in undesirable integrated change of other goal factors in a vector of goals.

Analysis of coordinates of a vector of control actions on consistency with a vector of goals. The concept of an EFC is inapplicable for control factors, as the analyst initially (before realization of the structure and goal analysis) does not know how the change of one or another control factor will affect the integrated behaviour of goal factors.

Definition 4. Vector of control factors is consistent with a vector of goals Y, if for each coordinate of a vector of control actions $U = (u_1 ... u_p)$ it is possible to determine such sign, that for a resulting sign vector *sign (U)* it will be fair:

$$r_s = sign(q_{st}) sign(u_t) \text{ for any } u_t \in U, y_s \in Y \qquad (4)$$

When control factors are consistent with the vector of goals and (4) is fulfilled, any change of control factors according to a vector *sign (U)* will not cause the change of any coordinate of a vector of goals Y in undesirable direction. Let $U^*(0)$ be a vector of control actions the signs of which are selected according to (4), and $|U^*(0)|$ - vector $U^*(0)$ in which all coordinates are replaced with their absolute values. The concepts entered above allow to formulate the following statement.

Statement. If the selected vector of goals Y is consistent and the set of control factors is coordinated with a vector of goals it is possible to choose such vector of control actions U for which it will be fair

$$\left|U^*_1(0)\right| \le \left|U^*_2(0)\right| \to Y\left(U^*_1(0)\right) \le Y\left(U^*_2(0)\right), \qquad (5)$$

where $Y\left(U^*_i(0)\right)$ is a vector of changes of goal factors caused by activation of vector of control actions $U^*_i(0), i = (1...m)$, i.e. property of "domination" by modules of control transfers into property of "domination" by results of their influence on goal factors.

In other words, more "intensive" control (with large absolute values of coordinates) will cause more "intensive" changes of coordinates of the goal vector in desirable directions.

The mentioned definitions are used in situation analysis and modelling. Thus, violation of conditions of consistency of the selected vector of goals can help the analyst to understand the interaction of goal factors and set his vector of goals "more correctly", conforming to the situation. Analysis of the vector of selected control factors on consistency with a vector of goals will allow to resign inconsistent control actions and, on the contrary, to more actively use "advantageous" control factors, the change of which according to control actions affecting them will result in great favourable changes of goal factors.

Analysis of efficiency of integrated influence of control factors on goal factors. The essence of realization of control actions consists in such change of control factors that their influence on goal factors will result in favourable changes of goal factors, i.e. to changes of goal factors in the direction of their EFCs.

In this connection it is important to answer the question "which of the control factors are the most "effective" for reception of a positive effect if compared by their integrated influence on goal factors?".

Formally, the parameter of efficiency $E(u_k)$ of the control factor u_k (i.e. the maximal positive effect from the change of u_k) is determined as absolute value of the sum of coefficients of influence of the given control factor u_k on the goal factors multiplied by EFCs of the goal factors, i.e.

$$E(u_k) = \left| \sum_{i=1}^{m} r_i q_{ki} \right|,$$

where r_i is the EFC of the goal factor y_i,

q_{ki} - (k,i)-th element of matrix Q.

Really, the maximal positive effect Δy from realization of control u_k on the factor x_k is estimated as

$$\Delta y = \left(\sum_{i=1}^{m} r_i q_{ki} \right) u_k,$$

where the sign of action g_k coincides with the sign of the sum

$$\sum_{i=1}^{m} r_i q_{ki},$$

and its value is equal to 1.

The above technique of structure and goal analysis as well as the technology of cognitive analysis and modelling of a SEO purposeful development (Maximov V., Kornoushenko E., Makarenko D., 2001) is supported by a dialogue software package (DSP) "Situation-2".

DSP "Situation-2" ensures:

1. Construction of cognitive model of a situation :

   - selection and substantiation of the basic factors of a situation;

   - establishment and substantiation of correlation of the factors;

   - construction of graph model of a situation.

2. Structural interpretation of problems requiring solution in the situation.

3. Structural analysis of the situation under research.

4. Searching and substantiation of strategies of goal reaching in stable or changing situations:

   - choice and substantiation of the desirable goals;

   - choice of activities (controls) necessary for reaching of goals;

   - analysis of basic possibility of reaching of goals from an initial state of a situation with the use of selected activities;

   - analysis of restrictions on a possibility of realisation of the selected activities in reality;

   - analysis and substantiation of a real possibility of goal reaching;

   - development and comparison of strategy of goal reaching.

5. Substantiation of possible scenarios of the situation development.

6. Machine generation of the reports and systematisation of results of a problem modelling.

DSP "Situation" allows to examine two classes of analytical problems:

- structural analysis of a situation;

- scenario exploration of a situation development trends.

DSP "Situation-2" was successfully applied for modelling of the control of strategic development of regions, markets, companies, and other problems.

REFERENCES

Maximov V. (2001). Cognitive Analysis and Situation Modelling. *Proceedings of the 8th IFAC Conference on " Social Stability: The Challenge of Technology Development " (SWIIS'01). Sept. 27 - 29, 2001. Vienna, Austria.*

Maximov V., Kornoushenko E. (2001). Analitical Basics of Construction the Graph and Computer Models for Complicated Situations *Proceedings of the 10th IFAC Symposium on Information Control Problems in Manufacturing (INCOM 2001). Sept. 20-22, 2001. Vienna, Austria.*

Maximov V., Kornoushenko E., Makarenko D. (2001). Use of Cognitive Modelling for Analysis of Socio-Economic Processes and Estimation of Variants of the Regional Development *Proceedings of the 10th IFAC Symposium on Information Control Problems in Manufacturing (INCOM 2001). Sept. 20-22, 2001. Vienna, Austria.*

ELSEVIER

IFAC
PUBLICATIONS
www.elsevier.com/locate/ifac

ANAESTHETISING OURSELVES: ENGINEERING AND TECHNOLOGY EDUCATION AS A BARRIER TO AN ETHICAL TECHNOLOGY PROGRAMME

L. Stapleton and C. O'Dowd Smyth

Waterford Institute of Technology
Republic of Ireland

Abstract: This paper proposes the idea of 'engineering consent' as an important ethical consideration for engineers. The paper illustrates the notion of techno-culture, emphasising the non-neutrality of technology in the world and how technology can be used in cultural and economic colonialisation by the west. It advocates a fundamental review of engineering education theory and practice. In this review, the paper argues that engineers must enter new spaces of thought and learning including the post-colonial 'Third Space' advocated by Homi K. Bhaba. The review of education must be based on current ideas as to professional competence, as well as a healthy approach to dissidence through innovative and creative thought processes. In this way a new community of practice will emerge which is centred not upon technological progress but social responsibility. *Copyright © 2003 IFAC*

Keywords: Engineering theory, ethics, education, social impact, social stability, reflective learning.

1. INTRODUCTION: ENGINEERING CONSENT IN AN OVERDEVELOPED WORLD

Many commentators on the nature of democratic societies in, so called, over-developed countries have pointed to the ways in which mass consent is manufactured (Herman & Chomsky, 1989; Chomsky 1994). 'Over-developed countries' is a recently coined phrase used to indicate those economies which are contributing to serious environmental problems as well as socio-economic disparity whilst maintaining strong economic growth going forward. They are typically western economies and the most stark example is the United States where a minority of the global population consume a majority of resources. The over-developed economies are in direct contrast to those economies, in places such as Africa, where a minority of the earth's resources are supporting a huge population and where, due to political and other structures, the situation is unlikely to be addressed in the near term.

The engineered consent referred to above is manifested in a narrowing of debate to reasonably 'safe' frames of reference, from the point of view of the power elites in those states. This is primarily achieved by a careful and subtle control of the media and education. Through this process societies become anaesthetized and therefore acquiescent. This maintains the 'stability' of democratic societies and ensures that the power elites are not threatened (Chomsky, 1994). In fact it has been stringently argued that this manufacturing of consent has been counter-productive leading to events such as September the eleventh 2001, in New York. In this view, it is argued that people are not allowed access to important debates so that they have little sense of the 'other' i.e. those outside my immediate community who may have a grievance. In this view, the atrocities of September 11 in New York were an inevitable consequence of foreign policies which were not challenged within the internal power structures of western states (Chomsky, 2001; Virilio, 1998). It is only now, in the aftermath of the events in New York and Washington, as the west dismantles Iraq and

Afghanistan, that many of the policies referred to by dissident commentators are being reviewed in more mainstream debate.

2. THE NON-NEUTRALITY OF TECHNOLOGY & COLONIAL STRUCTURES

Philosophers and researchers recognise that technology, in particular information technologies, telecommunications and televisual technologies, are cultural artefacts. As such they are non-passive and not culturally neutral. Instead they embody the culture from which they were derived and, consequently, some have called the transfer of technology across cultural barriers as the transfer of 'techno-culture' (Ihde, 1998). The technology-transfer process has been criticised because the cultural background of most information and televisual technologies, as well as the content portrayed through these media, resides in western, over-developed nations. Consequently, the export of technology and associated knowledge to other, non-western, cultures can be viewed as a new form of colonialism, structurally similar to the expansion of capitalism, where the current form of capital is, so-called, 'knowledge capital' (Stapleton , 2003). Indeed, it has been widely and consistently argued that technology is highly political, and culturally violent, in fundamental ways (Zizek, 2001; Bannerjee, 1999; Ezrahi, 1995; Rada, 1990; Arendt, 1970; Quoting Winner, 1988)

'we see.. an ongoing process in which scientific knowledge, technological invention and corporate profit reinforce each other in deeply entrenched patterns that bear the unmistakable stamp of political and economic power' (p. 27).

Echoing early political theorists like Machiavelli, two main social dimensions by which consent management is achieved are identified: education and the mass media (Chomsky, 1998; Skinner, 2000). It is self-evident that, according to these arguments, there are serious ethical implications where engineers and technologists who themselves are located within passive western societies, develop information technology and visual media which are exported across the globe. This process of cultural colonisation has been recognised and actively withstood by some western European societies where even language is in seen to be in danger of being subsumed into an Anglophone world.

2.1. Difficulties Resisting Mass Consent: The Case Of France

This perceived Anglo-Saxon linguistic and cultural hegemony is met with ongoing resistance in France. There, the elite watchdog committee the *Académie Française,* composed of French intellectuals, examines new terms and words of foreign origin and decides whether or not to allow these new words to be included in the French dictionary or to coin suitably 'French' terms instead. For example, the Internet was christened *'le toile'* but everybody in the French speaking world calls it *Internet,* with French pronunciation of course! It is perhaps interesting to note that with regard to internet use France is far behind other European countries. Also the French prefer to obtain their information from the television news rather than the print or electronic media. Images of French President Jacques Chirac addressing the nation in his newly-acquired role of *Résistant* to the Anglo-American war offensive against Iraq have made him a hero in the eyes of millions of French people. On the other hand, giant television networks such as CNN and Sky have ridiculed France's anti war (and therefore perceived anti-American) stance and have promoted the idea that France is the real enemy. As a result, sales of French goods have declined dramatically across the United States. However, contrary to a globalized perception of French people as being intrinsically anti the English language, a whole generation of highly educated 20 and 30 somethings, products of the elite Engineering and Business Schools, *Les Grandes Ecoles,* are preferring the use of English rather than French as a global tool of international communication. They perceive their government's linguistic isolationist policy as an impediment to growth and change. It can therefore be argued that access to the Internet, and its emphasis upon English, has changed the entire perception and outlook on the world of this young generation of French engineers and entrepreneurs.

This process, which is well documented elsewhere, highlights the link between, so-called 'Anglophone hegemony' and global economic activity. If France has difficulties withstanding what they perceive to be a cultural imperialism and invasion of their cultural space, then it is readily apparent that smaller, less affluent nations will experience severe difficulty resisting these pressures.

3. EDUCATION, ENGINEERING CONSENT AND SOCIAL STABILITY

The creation of a manufactured consent by which stability can be maintained raises important questions for those involved in the education of technologists and engineers. For example, is it possible that educators themselves can become complicit in this through the very process of education itself? The evidence suggests that modern western education actually ensures that those who pass through it are passive, non-questioning and acquiescent. This is especially true of engineering disciplines[1].

---

[1] Examples of this process are often cited in studies of ideology. For example, see Chomsky (1993) where he links intellectual dogma and ideology to human rights abuses.

Research in engineering education recognises the need for a more creative and innovation lead approach to teaching in this area. For example, changes in the role of engineers within society is leading to an evaluation of engineering education, which, in turn leading to a cry for the integration of creativity and innovative learning into engineering education programmes (van der Vorst, 1998)). This debate has highlighted the social responsibilities of engineers and implies the need for more creative thinking, and critical thinking, within the discipline. This has been further emphasised as a major issue for furthering research and practice in engineering ethics (van der Vorst,1998)). Brandt (1996), Juric et. al. (1999) and Acar (1998) highlight the international nature of this issue citing similar experiences and imperatives in Germany, Slovenia and the USA.

3.1.Re-Evaluating Engineering Education

Engineered consent is achieved through underlying deep-structures and processes within society rather than explicit and conscious policy. These structures become institutionalised and buried in the deep workings of societies and are, therefore, difficult to challenge or even expose (Foucault (1965), (1980), Kuhn (1996)). We shall now briefly review typical evaluation mechanisms used to assess engineering education programmes, highlighting how they can serve to maintain these structures.

The need for evaluation mechanisms is highlighted in the engineering education literature as means by which quality can be assured and educational standards maintained. These evaluation models and frameworks are to be used to highlight the effectiveness of engineering curricula from both the students and the teachers point-of-view. Whilst these are very useful devices, the methods rarely address deep structure issues associated with the particular curriculum. Consequently, the essential ethos of a course can remain untouched, and debate (or indeed dissent) concerning the essential values, which underpin the engineering subject, go unchallenged. For example, Atieh et. al. (1991) propose an excellent methodology for evaluating teaching effectiveness in engineering. It provides a sophisticated, multidimensional device for assessing perceptions of the effectiveness of engineering educational practice, utilising a mathematical model to deliver a score for individual courses. The variables utilised within the model at no stage address ethical content, openness of debate or independent learning. Instead, they focus upon a very narrowly defined set of input variables, including 'adherence to course syllabus'. Such course-evaluation devices cannot address deep structure issues within the education of engineers. Instead they are premised on the fact the overall educational process is appropriate in terms of creative thinking, reflection and so on.

4. THE NEED FOR CREATIVITY IN ENGINEERING CURRICULA

An emphasis upon dissidence and innovation has been shown to have practical application in the engineering of new products and technologies (Court (1998)). For example, Brandt & Ihsen (1998) illustrate how highly creative thinking applied to robotics creates entirely new research trajectories within engineering. Holmes (1998) shows that creativity in engineering thought is 'not an optional extra'. Instead, it goes to the very heart of engineering as a discipline, what we do and what we are as engineers. She illustrates how the development of a mature engineering ethics requires a fundamental shift in engineering thought towards, what she terms, the 'forebrain'. Platts (1998) shows how the development of technological artefacts goes hand in hand with the development of a sensitivity of who the others in our society are and how we can live together. He argues that 'the inter-twining of technical and moral creative skills' is central to engineering advancement, and indeed to the creation of civilised society. Lenschow (1998) argues for a paradigm shift in engineering education from 'teaching to learning', and he illustrates several ways in which this can be achieved.

It is readily apparent that, in spite of several innovative approaches to engineering education in isolated academic faculties, the education of engineers remains largely traditional. This tradition leaves untouched questions about the role of engineers and technologies within society: who gains and who loses in the race for technological progress. Neither does traditional education practice in engineering incorporate broader definitions of competence now appearing in the education research literature and recognised as important in other practice- oriented disciplines such as nursing and management (Carlile (2001), Golding & Currie (2000), Burnard (1995)). The idea of engineering education quality remains fixed upon criteria established by quality standards such as ISO, rather than on fundamental advances within educational philosophy and practice. Engineers from traditional engineering programmes define the quality standards used to measure advances in engineering education. This is the essence of manufactured consent and typifies how Foucauldian power structures are reinforced in society (Kuhn (1996), Foucault (1965)).

4.1. Experiences of a Language Lecturer Amongst Engineering UnderGraduates

One of the authors currently lectures in European languages to a variety of engineering undergraduate classes: in this case compulsory foreign language teaching to Degree in Bachelor of Engineering, Bachelor of Engineering Technology and Bachelor in Manufacturing Systems students. Her views and experiences shed some light on the way in which engineers approach education, as compared to the

Humanities. She firstly notes an emphasis on technology as the principal means of disseminating information. In her experience, an over reliance on PowerPoint presentations and overheads had rendered the engineering student passive and incapable of independent enquiry and thought. The students tended not to understand the relevance of learning a foreign language and they were not educated as to its importance. The students expected task-based learning exercises (such as grammar exercises, translation exercises etc) and were horrified when the communicative approach was promoted. They seemed not to be able to make the connection that language is principally a means of communication, of *active communication*. Autonomous learning does not seem to be part of their repertoire. However, when reflective learning and autonomous learning strategies were introduced into the learning process, students became empowered with the sense of their learning of a language as a self-determined process, with the lecturer as facilitator and not director of that process. At the end of the year the positive response of engineers to this approach is very apparent and they describe being enabled to think for themselves and assess their own needs, thoughts and ideas. The other author introduced a similar, independent-thought oriented, approach amongst fourth year computing students in an information systems course. Whilst students took a little while to adjust a questionnaire handed out at the end of the year registered a 91% satisfaction rating with the alternative approach when compared to the more traditional approach. Students told how they were able to learn much more by engaging in research, thinking through the basic assumptions of their discipline and developing more critical faculties in the complex area of information systems.

5. FROM ENGINEERING CONSENT TO ENGINEERED DISSIDENCE

As illustrated in the opening section of this paper, it is readily apparent from the political literature that many academic institutions still provide an important means by which power elites within society can be maintained and strengthened. In order to combat this within engineering, engineers and educationalists need to reconfigure educational processes to allow for increased freedom of debate, through an emphasis upon the engineer as located within society – rather than the engineer simply located within a profession. This requires new ways of thinking about professional development in engineering. These new approaches must be based on advanced, broadly-based models of engineering competence as hinted at by engineering educationalists elsewhere but rarely addressed in the literature (see for example Brandt (1996). We need to adopt recently developed notions of professional competence such as Carlile (2001). In his work he states that the four main dimensions of competence include Administrative Competence, Technical

Competence, Personal Competence and Ethical Competence. The latter two refer to life-skills and moral judgement and incorporate these as essential dimensions of any programme wishing to address issues of competence. In order to enable engineers to enter new spaces of competence and learning it is appropriate to briefly map out recent developments in post-colonial criticism and assess how these developments can be utilised to transform key aspects of engineering education and thought.

5.1 A Post-colonialist Approach to Engineering Education: The' Third Space'

In order to overcome deep structural problems within engineering education, and thus encourage free thought and dissidence, we need to rethink the approach we take in thinking about culture in engineering. Typically, engineering ethics debates, like many other cultural debates, focus upon binaries i.e. 'me' as opposed to the 'Other'. This is reflected in notions such as the 'engineer'/'user' binary for example. These binaries are often associated with cultural hegemonies and identified in the deep structures of cultural discourse. The postcolonial critic Homi K. Bhaba has dismissed the notion of fixed binary definitions in relation to hegemonic discourse such as: teacher/student, master/servant, native/foreigner in favour of a new 'art of the present'. This approach requires us to 'think beyond narratives of originary & initial subjectivities' and focus upon the *processes* produced in the articulation of cultural differences. He advocates a hybrid or 'Third Space' as a place of possibility and an agency for new ideas (Bhaba (1994)). In this process-oriented view, the subject is empowered to intervene *actively* in the transmission of cultural inheritance or transition, rather than *passively* accepting transmitted customs and pedagogical wisdom. He or she can question, refashion, or mobilise received ideas (McLeod (2000). Such an approach actively encourages interventions in technology design by all, and opens a way for a new discourse about what technology design should be all about. In this view, the engineering student is empowered to act as an agent of change, deploying received knowledge in the present and *transforming it as a consequence.* Thus received wisdom (as embodied in engineering syllabi) becomes a starting point for debate and criticism, rather than an end in itself. Deconstruction is a useful tool in promoting a non-hegemonic process. It creates an energy that decentres the cultural centres that represent former (and present) colonial networks, and contributes to the recognition of new forms of thinking from non-hegemonic sources (Orlando, (1999).

In this process difference is no longer a negative (Deleuze, 1968). Instead it functions as a means of *affirmation*. Representation is 'replaced by the expression or actualisation of ideas, where this is understood in terms of the complex notion of different/ciation'. For Deleuze, modern thought is the product of the failure of representation or the loss of identities and of the discovery of all the forces that act under the representation of cultural similarities. Those who do not fit some hegemonistic view of 'normality' are banished to the fringes of cultural life. Our modern world is thus simply one of simulacra in which repetition plays upon repetitions and differences play upon differences (Baudrillard, 1994). Quoting Bell Hooks (1990)

"To imagine is to begin the process that transforms reality".

6. CONCLUSION

It is self-evident that a reflective-learning based model of technological and engineering education is urgently needed and should be widely adopted. It refutes the idea that technology itself will, by its very nature, overcome the traditional limitations of western educational systems (Dreyfuss, 2001). Instead it argues that we need advanced educational processes and interventions to ensure that technology does not become another nail in the coffin of free-thought. The arguments presented here ensure that students are made aware of the particular context in which they operate, and can critically judge the processes and products they develop. Furthermore, this educational model demands a research lead approach, and can create great intellectual challenges for the educationalist herself. This paper argues that unless society addresses engineering and technology education as a process of socialisation, ethical demands cannot be satisfactorily fulfilled. Only by revisiting fundamental practice in engineering education can a new, dissident and open engineering education process be established by which engineers not only concern themselves with ethical issues, but lead the way in ensuring that communities and individuals alike are empowered by advances in technology. Even Machiavelli promoted dissension as a means by which the power of elites within republics is curbed (Skinner, 2000). This model of engineering education can help contribute to the counter-balance desperately needed if we are to overcome the engineering of consent.

ACKNOWLEDGEMENTS

The authors wish to acknowledge the helpful contributions and comments of the reviewers. They also want to particularly thank Professor Dietrich Brandt in helping put together the final paper.

REFERENCES

Acar, B. S. (1998). Releasing Creativity in an Interdisciplinary Systems Engineering Course', *European Journal of Engineering Education*, 23, 2, pp. 133-140.

Arendt, H. (1970). *On Revolution*, Viking Compass: NY.

Banerjee, R. (2001). Biodiversity, Biotechnology & Intellectual Property Rights: Unpacking the Violence of 'Sustainable Development', 19th Standing Conference of Organisational Symbolism (SCOS XIX), Dublin, (forthcoming).

Baudrillard, J. (1994). *Simulacra and Simulation*, Michigan University Press: Mi.

Baudrillard, J. (1999). *The Consumer Society: Myths and Structures*, Sage: NY.

Hooks, B. (1990). *Yearning: Race, Gender, and Cultural Politics*, South End Press

Bhaba, Homi K. (1994) *The Location of Culture*, Routledge, London & New York.

Brandt, D. (1996). Patterns and Challenges of Unergraduate Project Work in Germany: The Aachen Experience', *European Journal of Engineering Education*, 21, 2, pp.197-204.

Brandt D. & Ihsen, S. (1998). Creativity: How to Educate and Train Engineers, or Robots Riding Bicycles', *European Journal of Engineering Education*, 23, 2, pp. 131-132.

Burnard, P. (1995). *An Experiential and Reflective Guide for Nurses*, 3rd Ed. Butterworth Heinemann: UK

Carlile, O. (2001). Incompetent Teachers in Irish Voluntary Secondary Schools: Principles' Assessments, Attitudes and Reactions, Unpublished Ph.D. Thesis, University of Hull.

Chomsky (1993). *Year 501: The Conquest Continues*, Verso.

Chomsky, J. (1994). *Keeping the Rabble in Line*, Verso.

Chomsky, J. (1998). *The Common Good*, Odonian Press: Canada.

Court, A. (1998). 'Improving Creativity in Engineering Design Education', *European Journal of Engineering Education*, 23, 2, pp.141-154.

Deleuze, Giles (1968). *Difference & Repetition*, Cambridge Univ Press (Trans. Paul Patton, 1994).

Dreyfuss, H. (2001). *On the Internet*, Routledge: London.

Ezrahi, Y. (1995). Technology and the Illsuion of the Escape from Politics, in Ezrahi, Y., Mendelsohn, E. & Segal, H. (eds.), *Technology, Pessimism and Post-Modernism*, University of Massachusetts, MA. pp. 29-38.

Foucault, M. (1965), *Madness and Civilisation: A History of Insanity in an Age of Reason*, Vintage: NY.

Foucault, M. (1980). *Power and Knowledge: Selected Writings*, Pantheon: NY.

Golding, D. & Currie, E. (2000). *Thinking About Management: A Reflective Practice Approach*.

Gutierrez, C. (eds.), *Computers, Ethics and Society*, Oxford: UK, pp. 262-277.

Herman, E. & Chomsky, N. (1988). *Manufacturing Consent*, Pantheon.

Holmes, S. (1998). 'There must be More to Life Than This', *European Journal of Engineering Education*, **23**, 2, pp.191-198.

Ihde, D. (1998). *Expanding Hermeneutics: Visualism in Science,* Northwestern University Press: Ill..

Kuhn, T. (1996). The Structure of Scientific Revolutions, 3rd Ed., University of Chicago Press.

Mc Leod, John. (2000) *Beginning Postcolonialism*, Manchester University Press, Manchester & New York.

Orlando, Valerie. (1999) Nomadic Voices of Exile: Feminine Identity in Francophone Literature of the Maghreb Ohio Univ Pr

Platts, J. (1998). 'Participating in the Work of Creation', *European Journal of Engineering Education*, **23**, 2, pp.163-170.

Rada, J. (1990). *Information Technology and the Third World*, pp. in Erman, M., Williams, M &

Skinner, Q. (2000). *Machiavelli*, Oxford University Press: Oxford.

Stapleton , L. & Murphy, C. (2003). Revisiting the Nature of Information Systems: The Urgent Need for a Crisis in IS Theoretical Discourse, *International Transactions of Information Systems*, forthcoming.

Virilio, P. (1998). *The Art of the Motor*, University of Minnesota Press: MI.

Winner, L. (1986). The Whale and the Reactor: A Search for Limits in an Age of High Technology, University of Chicago Press: Chicago.

Zizek, S. (2001), On Belief, Routledge: London.

ELSEVIER
IFAC
PUBLICATIONS
www.elsevier.com/locate/ifac

A GUIDE TO MANAGE CONFLICTS DURING THE IMPLEMENTATION OF IT SUPPORT TOOLS

S. Schweditsch

IBM Austria
Obere Donaustr. 95, A-1020 Vienna
Sascha_schweditsch@at.ibm.com

A. Pérez Alonso

FH JOANNEUM Gesellschaft mbH
Werk VI Str 46, A-8605 Kapfenberg
azucena.perez-alonso@fh-joanneum.at

Abstract: In search for competitive advantages companies permanently strive for productivity improvements through IT solutions. Albeit selecting the most appropriate IT system is critical the themes that make or break the project's success gravitate towards the human side of an organization: change -, conflict management and dealing effectively with the predominant corporate culture. This paper deals specifically with managing conflicts between corporate stakeholders and employee resistance against changes in work processes brought about by ERP-systems. *Copyright © 2003 IFAC*

Keywords: Business Process Reengineering, Management, Implementation, Human Factors, Behaviour

1. INTRODUCTION

In this paper the authors first define key phases and concepts associated with the implementation of ERP-systems. They then go on to elaborate on cultural challenges and the relationship between change and conflict. A practical system of classification and typology of changes is offered. The paper will be closed with a case study of and Austrian manufacturing company which successfully implemented an ERP system applying the tools and methodology offered in this paper.

2. DEFINITION OF TERMS

1.1 Implementation

Can be defined as all of the organizational activities working toward the adoption, management, and routinization of an innovation (Tornatsky et al., 1983):
In the literature of implementation there are three different approaches (Loudon& Loudon 1996)

- Some researches focus on actors and roles. The idea behind it is that the organization selects employees (actors), which should develop organizational roles in order to innovate successfully. This literature focuses on adoption and management of innovations.
- A second approach emphasizes strategy. The spectrum of economic innovation can be thought of as two extremes on a continuum: the top-down innovation and the grass-roots innovation. This literature focuses on strategies of innovation.
- The last approach accentuates organizational change. In order to successfully manage change in the long term, organizational aspects must be in foreground (Yin, 1981). This third approach focuses on organizational change.
The term implementation will be applied according to the third approach of organizational change, accentuating organizational aspects.

2.1 ERP-system, enterprise resource planning

An ERP-system is an enterprise wide package that tightly integrates all necessary business functions into a single system with a shared database (Lee & Lee, 2000)

The ERP concept can be viewed from a variety of perspectives:

- ERP as a commodity, computer software
- ERP as a development objective of mapping all processes and data of an enterprise into a comprehensive integrative structure
- ERP as key element of an infrastructure that delivers a solution to business (Klaus et al., 2000)

The authors refer to ERP systems in the sense of a tool to integrate processes and data since the first definition is too narrow due to the fact that ERP in today's business is at the core of a company's structure, communication and processes. The authors agree with the definition given by Adam & O'Doherty, 2000: ERP systems centralized the information required by the company in one single system providing organizational actors with a common pool of data.

ERP systems are generally adopted to improve the company's productivity and it is contemplated more as a business solution than an IT solution (Kremmers and Dissel, 2000)

The origin of ERP is related to the concept of MRP (material requirement planning) systems, which were developed to calculate the necessity of materials (Klaus et al.; 2000). In the 70's MRP developed into MRP II (manufacturing resource planning) systems and included other modules such as Finance and Accounting or Human Resources. It was Gartner Group who created the term "ERP" in the 90's. By the late 90's companies were spending over $ 23 billion a year on ERP systems and at least 30.000 companies had implemented ERP systems (Merbert et al., 2003).

The company in the case study which follows chose a SAP ERP solution. SAP was founded in 1972; SAP is the recognized leader in providing collaborative business solutions for all types of industries and for every major market.
Headquartered in Walldorf, Germany, SAP is the world's largest inter-enterprise software company, and the world's third-largest independent software supplier overall. SAP employs over 28,900 people in more than 50 countries. SAP counts with 12 Million Users, 60,100 installations. 1,500 partners, and 23 industry solutions.

3. CHANGE MANAGEMENT

Nothing endures but change.
(Heraclitus fl.c.500 BC)

Change Management is a systematic approach to dealing with change from the perspective of an organization and at the individual level. For an organization, change management means defining and implementing procedures and /or technologies to deal with changes in the business environment and to profit from changing opportunities (Kotska & Mönch, 2002).

Following the Webster's Ninth New Collegiate Dictionary: " Change" is:

- To make different in some particular
- To make radically different
- To give a different position, course, or direction to
- To replace with another
- To make a shift from one to another
- To exchange for an equivalent sum or comparable item
- To undergo a modification of
- To undergo transformation, transition or sustition
- SYN: alter, vary, modify

" Manage" is defined as:
- To handle or direct with a degree of skills or address
- To treat with care
- To exercise executive, administrative and supervisory direction of

The authors will use the term Change Management in the context of this paper in the sense of the following definition "The direction with a degree of skills of innovations which will give a different position, course or direction to the processes of a company "

A **Change agent** is the person who develops technical solutions but also redefines the configurations, interactions, job activities and relationships of various organizational groups; he/she is the catalyst during the change process to ensure the adaptation to a new system or innovation

Change management can be contemplated as the convergence of two fields: the engineer's field of improving business performance and the psychologist's field of managing the human side of change (Jeff Hiatt & Tim Creasy, 2002).

The mechanical focus of change is central to make changes to the operation of a business; this perspective focuses on observable, measurable elements that can be changed or improved. From this perspective a business is a mechanical artefact where pieces can be modified, redeployed or replaced to reach the desired solution. Companies focusing exclusively on this approach will have to deal at some point in the process with the issue of human resistance to change.

The psychologist's field focus on how employees behave when being confronted with a changing environment and how they can be positively involved in the change process.

The table 1 below summarizes and contrasts the two approaches:

	Engineer	Psychologist
Focus	Processes, systems, structure	People
Business practices	BPR, ERP, TQM	Human Resources, OD
Starting point	Business issues or opportunities	Personal change/ employees resistance
Measure of success	Business performance, financial and statistical metrics	Job satisfaction, turnover, productivity, loss
Perspective of change	" Shoot the stragglers, carry the wounded"	" Help individuals make sense of what the change means to them"

No single approach serves as panacea as the authors will show in the last part of the paper. The convergence of the two approaches is essential, though, to implement change successfully. Performance, strategy, processes and systems are key issues to consider when deciding the direction of change. However an organization must also contemplate the implications of any change for its culture and values and take into account the capacity of change of its employees. In addition the psychological field must be dealt with in order to be able to introduce change on the mechanical side.

There are two types of change (Osterhold, 2002) to be considered:
- Transactional changes are improvements of the existence structures, without putting in question the values and behaviour patterns
- Transformational change concerns a company's strategy, technology, structure and personnel decisions. Values, paradigm and behaviours will be questioned.

It depends on the company, on the corporate culture, on the compatibility with IT systems, if the implementation of ERP systems presupposes a transactional or a transformational change; usually companies expect a transactional change and are surprised by a transformational change which overloads the capacities to manage it. This is because the psychologist focus must be considered with the same importance and weight as the engineering's one.

The traditional values that have been the centre of traditional organizations: control, consistency and predictability (Peter Block) have passed through a progressive evolution. The new values in companies include empowerment, accountability and continuous process improvements. The changing landscape of corporate values makes business changes all the more difficult and turns them into a greater challenge.

Research with of more than 320 projects showed that the fundamental reason for failure in change processes is the inability to manage the change. Failing to manage the human side of change results in inefficient, expensive and unsuccessful projects, job dissatisfaction and high turnover.

4. MANAGING CONFLICT DUE TO CHANGES CAUSED BY THE IMPLEMENTATION OF ERP SYSTEMS

One of the essentials points to manage the change is to have the ability to manage the emerging conflict occurring during the implementation of changes.

Every change goes though different phases of acceptance (curve of acceptance of change processes). The change agent in the company should take care that the phase of shock and denial go as smoothly and quickly as possible trying to move to the learning phase fast by offering support:

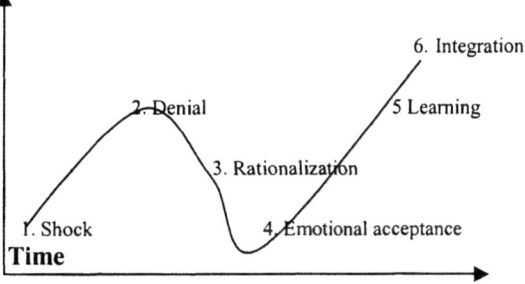

The graphic shows that the rational acceptance of a system does not mean that the employees are ready to learn and use the new programs; an additional step is required: the emotional acceptance of the change, the users have to feel comfortable with the new situation, and companies must understand and facilitate it. In change processes it is important to recognize that employees – and not management – must implement changes at the operative level. The first rational - and then the emotional disposition to accept the changes will decide the success of the implementation. The change is fully implemented as soon as the perceived competence reaches the point of integration (see point 6 in the chart above). It is at this point that the new behaviour is integrated into daily routines and that a fusion between the old experiences and new business processes has occurred (Senge, 2000) – old knowledge and newly learned skills have merged.

Webster's Dictionary describes conflict as "a battle, contest of opposing forces, discord, antagonism existing between primitive desires and instincts and moral, religious, or ethical ideals." Conflict occurs when two or more people oppose one another because their needs, wants, goals or, values are different.

Conflict management is the practice of identifying and handling conflict in a sensible, fair, and efficient manner.

But why should employees resist change? What are the reasons for conflicts? The literature gives us an endless list of factors, which determine the success of the implementation, the most accepted are the following (Patzak/ Rattay, 1998):

- The goal of the projects is not clear; top management decides what to implement without consulting with employees.
- The goals of the project are clear, but not accepted - The roles and duties are not clearly defined, competencies are unclear.
- Tayloristic project work. The initiator team acts very active and differentiates itself more and more from other employees, so that at the end the implementation is seen as the initiators' exclusive project. The project will be modularised.
- The new competencies presuppose more information, time and abilities than employees have available.
- Lack of project management skills
- Isolation of the ERP System from other parallel processes
- Isolation of the union
- Change in the structure of the jobs, rebuilding processes,
- Conflict of personalities, fears of new situations and distrust of ones own capabilities.
- The implementation is not supported strongly enough by top management.
- Unrealistic planning of resources

The survey of Deloitte & Touche 1993 of 400 US and Canadian CIO´s show us the biggest obstacles to business reengineering success (note total exceeds 100% due to multiple answers).
Resistance to change : 60%. Limitation of existing systems: 40%. Lack of executive consensus: 38%. Lack of senior executive champion 37%. Unrealistic expectations 28%. Lack of cross-functional project team 28%. Inadequate team skills26%. IS staff involved to late 17%. Project chapter too narrow 13%.

After reviewing the causes of non acceptance and the indicators for a successful implementation the authors proceed to concretise the critical aspects for the implementation of SAP (Blume, 97).

- Mission Statement and goals must be defined from the beginning. Stabilizing a mission can make complex relations and dependences clearer and promote the idea of a team
- Management and Union must manage the reorganization process together at the strategic level
- Integration of work plan, organization and ERP training; labour conditions must be adapted to allow he technical implementation. There is no possibility for progress and advance if the company does not support a learning organization which supports employees development and learning and which give time to learn and for a learning transfer to occur.
- Constructive implementation of social responsibilities according to the law. Regulations in reference to data protection, health regulation, equal treatment etc. must be taken into consideration before beginning with the design of the process and technical implementation
- Involvement of the employees. Employees should not only be asked about the know how of their work place but about desires and suggestions to design the processes due to the huge implication it can have for the new job design. In this way it will be possible to satisfy employees and to motivate them to put their knowledge to service for the new systems.
- Cautious integration of processes and reorganization trough the system. Technical integration per se should not be a goal; in some cases the overload of data and the access to information can retard the process flow. The integration must be rational and step-by-step.
- Realistic combination of reorganization and ERP- implementation. Technical and organizational aspects should be run parallel and compatibly.

The structures of the organization must be able to absorb the technical implementation; in negative case the management should first prepare the structures, processes and company culture for the changes ahead.
Without doubt the implementation of ERP systems contributes to the evolution (in some cases revolution) of the structures and processes of the organizations. The traditional dichotomy (centralization versus decentralization) gains in importance again by large scale ERP projects (Monteiro, Eric, HepsØ, 2002). From the socio-technological point of view ERP can be contemplated from two very different perspectives: as instrument of control by offering visibility and transparency (centralization instrument). As Davenport (1998) writes: we "need to understand the problem (ERP systems) are designed to solve: the fragmentation of information in large business organizations", the emphasis put on the formation and circulation of "best practices" contributes to a

uniformity, standardization, modularisation and specialization which always supposes a centralization of information and processes. On the other hand ERP systems can be consider as a response to problems related to hierarchical organizations offering cross-functional cooperation and a more process-oriented approach and as consequence as an instrument of decentralization " SAP R/3 overcomes the limitations of traditional hierarchical and function-oriented structures like no other software. The functions are integrated into a workflow of business events and processes across departments and functional areas (www.sap.com) as a facilitator of empowerment among the employees, in every case offers an opportunity to redistribute the power in the company (decentralization instrument): The technological architecture offers few main-frames serving a community of connected terminals while the diffusion of networks can be contemplate as a decentralizing effect (Monteiro, Eric and Hepsø, 2002). However the discussion is not closed and it is dependent on the culture, organization, values and structures of the company it can be used as a control or an empowerment instrument, since technology should assist the company and be adapted to the characteristics of the company and not vice versa.

There is an interaction between: mental models, technology and social structures (like rewards and incentives) and only when the three of them have been harmonized we can speak about a successful implementation (Orlikowski, 1992). Finally, it is important to considerer these four statements to understand the relation between human beings and technology:
- Technology is the product of human action
- Technology is the medium of human action
- Institutional conditions affect interaction with technologies
- Human interaction with technology affects the properties of an organization

5. THE IMPLEMENTATION OF SAP R/ 3 IN AN AUSTRIAN COMPANY

The constellations of the project exposed in this paper - even if quite complicated - is not unusual due to the complexity and dimensions of the project(Wildbacher, 2001)

The client is a manufacturing company. The company has a business volume of 43 million Euros. The core business is the production of bands for transport. The export sales are 97% and the company has 220 employees. The structure and organization of the company can be considered flat. The goals of the implementation are: the reduction of redundancy, paper flow, errors, the increase of flexibility and business volume. The implementation of SAP was only a part of the desire to reorganize but at the end due to the emerged problems took the central role of the reorganization.

The solution provider: It is a large company with more that 33.500 employees worldwide, it is leader in electronic solutions.

The consultant: It is a small company, which offers support in reengineering and conflict situations. Its mission in this project was first to analyse the processes of the client, to propose reorganization measures and to coordinate these reorganization measures with the implementation of SAP R/3.
The division of roles was as follow:
- Executive committee (two persons of the client company, the senior consultant, manager of the IT solution provider)
- Project manager from the client + Assistant
- Project group: Key user + users + module responsible from the solution provider company +a member of the consultant company

The members of the consultancy company were assigned to different modules as support and bridge between key users and solutions provider's employees. Furthermore they had another meta-function which was to coordinate the implementation with other measures in the reorganization.

The three parts of the constellation were involved from the conception phase until the end. At the beginning the consultant analysed the core company's processes: management customer orders, production and customer service also from the first moment they did a strengths and weakness analysis to be considered during the implementation. The project was divided in four phases: conception, implementation of the HR module, implementation of other modules and optimisation. The implementation was carefully planned. from the mechanical point of view but in an unrealistic and simplistic way. The problems emerged already in the second phase and after the comparison between the target and the actual situation it could be concluded that there were technological deficits in the functionality of the systems. The functionality of the whole system was never tested. From the organizational and human side cooperation across modules lacked. There were not enough personnel available and employees neither had the time nor the inclination to learn SAP. After the second phase the project was delayed more than six months and at the end employees began to see the implementation as an additional burden which interfered with their regular job.

A survey was designed to analyse the problem, the questionnaire was divided in 4 categories: time, contents, training and product. In brief, the biggest problem the employees saw was time and training. They were satisfied with the implementations content and SAP was generally accepted. The differences between departments were big being the sales departments the most problematic. But why was the project not a total success after so many plans and investments in coaching, training and in products? After all the client invested a lot of financial resources to make the implementation a success, they engaged a consultancy company who was supposed to take care of the human aspects of the change. Revising the success factors of Blume 1997 we can not conclude 100 % the causes of the failure but we can point to the following:

- They was no mission statement to communicate to the whole work force
- Management and Union did not work together and the union was not included in the management committee
- Integration of work plan, organization and personnel development in the ERP Project was tried but the employees had the feeling they must do the training parallel to their regular obligations. The emotional resistance was increasing with the time.
- Health regulations were not taken into consideration.
- The involvement of the employees for the conception was non-existent.
- The process took much time. The initial idea was a big reengineering of the company, but maybe it was too ambitious and quick and not adapted to the capability of change of the work force
- Realistic combination of reorganization and ERP- implementation. The implantation was faster than the reorganization and this caused many insecurities and resistances among the employees.

This little case shows us the complexity of the implementation, the company invested a lot of resources in time, in training but one aspect was neglected: in order to succeed there must be compatibility between IT solutions, structures of the company and mental models of the work force.

REFERENCES

Blume A (1997). Projektkompass SAP. Vieweg Business Computing

Ciborra, C (editor) (2000). *From control to drift. The dynamics of corporate information infrastructures*. Oxford University Press

Ciborra, C., & Lanzara G.F (1994). " Formative Contexts and Information Technology: Understanding the Dynamics of Innovation in Organizations". Accounting, Management and Information Technology

Davenport T.H (1998). Putting the enterprise into the enterprise system, *Harvard Business Review*, July-August, pp120-131

Hyatt J. & Creasy T (2002). The definition and History of Change Management. BPR Online Learning Center

Kostka C. & Mönch A (2002). Change Management. Hanser

Kremmers M. , Dissel II.V (2000). . ERP Systems migration Communications of the ACM. **Vol. 43**

Laudon & Laudon (1996). *Management information Systems*. Prentice Hall

Mabert V, Soni A and Venkataranan M. (2001). ERP: Common Myths versus Evolvine Reallity. Business Horizons

Monteiro, Eric and Hepsø. Implementing multi-site ERP projects: centralization and decentralization revisited. www.idi.ntnu.noSie

Monteiro, Eric and Hepsø, Vidar (2000). Infrastructure strategy formation: seize the day at Statoil. In C. Ciborra (ed): From control to drift. The dynamics of corporate information infrastructure, Oxford Univ. Press pp.71-84

Orlikowski W (1992). The duality of technology: Rethinking the concept of Technology in Organizations"

Osterhold Gisela (2000). Veränderungsmanagement. Falken und Gabler

Patzak G, Rattay G (1998) " Projektmanagement. *Leitfaden zum Management von Projekten, Projektportfolios und projektorientierten Unternehmen*".Linde

Tang M, Soh C, Boh, (2002). Enterprise resource planning (ERP) systems as a technology of power: empowerment or panoptic control?,. In Critical Analyses of ERP Systems: The macro level. ACM Press

Tornatsky L (1983). The Process of technological innovation: Reviewing the literature. Washington, D.C: National Science Foundation

Wildbacher Regina (2001). Projekt- und Konfliktmanagement bei der Einführung eines ERP- Systems

Yin R. (1981). "Life histories of innovations: How new practices become more routinized". Public Administration Review

The Ethics of Military Work:
What can be Learnt from The Application of Ethical Theories

M.A. Hersh,
Department of Electronics and Electrical Engineering,
University of Glasgow, Glasgow G12 8LT, Scotland.
Tel: +44 141 330 4906. Fax: +44 141 330 6004. Email: m.hersh@elec.gla.ac.uk

Abstract: This paper briefly presents some of the main theories of ethics and then applies them to the case of military research. The study illustrates the range of ethical issues associated with military research, as well as the limitations of each theory on its own. It also highlights the advantages of the method developed by the author of combining a number of different theories to gain a full picture of all the ethical issues. *Copyright © 2003 IFAC*

Keywords: Ethical theories, military research, type of funding

1. A BRIEF INTRODUCTION TO DIFFERENT ETHICAL THEORIES

The terms ethics and morals are frequently used interchangeably, but it can be useful to distinguish morality as concerned with right and wrong conduct and motives and ethics as the philosophical study of morality (Gluck, 1986). Thus ethics can be seen as a framework in which to study moral dilemmas (Bennet, 1996) and the ways in which they can be resolved (Vesilund, 1998). However, in practice, the term ethics is generally used to describe right and wrong conduct and motives in a professional context and will be used in this way here. In some cases legal and ethical obligations can come into conflict (Park, 1995), for instance with regards to the disclosure of information restricted by considerations of national security.

One approach to analysing ethical dilemmas involves the application of different ethical theories. One of the simplest categorisations of ethics is into consequentialist and deontological: consequentialist approaches are concerned with consequences and the balance between benefits and harms, whereas deontological ones also consider the intention and the innate virtue of a course of action. Ethical principles can also be classified as universalist or absolutist and

situation based. Absolutist approaches assume that a particular set of ethical principles is always valid, regardless of the surrounding circumstances, whereas situation-based ethics modify ethical principles or prioritise them differently to take account of the particular situation. Although in many ways more realistic, care has to be taken to ensure that the application of situation-based ethics is not used as an excuse to avoid hard ethical issues.

Other ethical theories include virtue ethics, utilarianism, duty ethics and rights ethics (Babcock, 1991; Madu, 1996; Martin et al, 1996). Utilarianism only considers consequences and that actions should result in the greatest good for the most people (and sometimes also animals), whereas duty ethics focuses on actions rather than consequences and is based on the idea of duties or responsibilities and respect for persons. Rights ethics considers actions to be wrong if they violate fundamental moral rights, whereas virtue ethics supports actions which build good character. Utilitarianism can be divided further into positive and negative utilitarianism (Lappé et al, 1999).   Positive utilitarianism assesses new technologies in terms of their benefits against the risks and costs and generally favours new technologies and pays little attention to the risks of, for instance, the destruction of ecosystems. Negative

29

utilarianism is mainly concerned with offsetting or mitigating present or future harms and is more obviously compatible with the precautionary principle (Agenda 21, 1992).

Normative ethics is based on the definition and defense of basic principles and virtues, such as beneficence, justice and autonomy. The application of these principles to specific ethical problems, including those arising in the professions, is referred to as applied ethics. Beneficience involves the active promotion of acts that benefit others, helping people to further their legitimate interests and removing or preventing possible harm. Justice involves behaving fairly and in accordance with what is owed or due. Distributive justice requires the just distribution of social benefits and burdens and equal treatment of all. However, unequal treatment may sometimes be required to alleviate structural or other inequalities and in this context should be considered just (Barbour, 1995).

Other less commonly used approaches to ethics include: the ethics of care consisting of a context based approach to preserving relationships (Gilligan,1982); the ethics of social experimentation in which engineering projects and the introduction of new technologies are considered as experiments (Martin et al, 1996) and ecocentred ethical approaches with a holistic perspective based on ecological systems (Callicott, 1992). The experimentation approach explicitly draws attention to the requirement for informed consent based on sufficient and appropriate information and voluntary participation, whereas ecocentred approaches focus on connections and interactions and therefore increase the likelihood of awareness of long-term and indirect consequences. Rule based ethics is based on the application of rules, such as the codes of ethics or professional conduct of science and engineering societies. Their provisions (Oldenquist et al, 1979) can be divided into three categories: public interest, desirable qualities and professional performance. However such codes rarely indicate how decisions should be made in the case of conflicting obligations, although it is such conflicting obligations that frequently gives rise to ethical problems.

Ethical theories generally recognise that individuals have duties and responsibilities to themselves as well as wider ones to society. However engineers and others who act ethically, for instance by 'whistle blowing' (Hersh, 2002a) to draw attention to potentially hazardous or unethical practices in their firms or through refusal to carry out work that they consider unethical, may suffer loss of employment or financial penalties. This individual jeopardy could be reduced by moves to more collective responsibility, and the development of an organisational or social culture of responsibility. Each of the theories presented above has both advantages and limitations. In particular there has been considerable criticism, as well as spirited defence of utilarianism (Schleffler, 1994), which is discussed further by the author in (Hersh 2002b). She also notes the serious drawbacks of the tendency to apply single ethical theories on their own to complex problems and has therefore developed a multi-criteria approach which combines a number of different ethical theories, analogously to the use of multi-criteria optimisation or decision making. This paper will apply the ethical theories discussed in this section to analysis of military research and civilian research with military funding.

2. MILITARY RESEARCH

This section contains a brief summary of some of the issues discussed by the author in (Hersh, 2001). Although the term 'defence' is commonly used with regard to military research and expenditure, an increasing proportion of armed conflicts are now within rather than between countries (Renner, 1993). In addition the richer and more influential countries have always had a preference for fighting their wars on other nations' territories. Thus the use of weapons by the richer countries, in which most of the high level military research takes place, has been outside their territories over the last fifty years and is in general more appropriately classified as offensive than defensive. A sizeable proportion of military production (from the richer 'west') is exported (Hartley et al. 1995) and controls over recipient countries and permitted uses, including internal repression, are generally limited. Changes in political alliances result in armaments sometimes being used against the forces of the country that sold them, as in the case of the weapons used by Argentina against UK forces in the Falklands conflict (Evans et al., 1991).

Thus the relevant ethical question may have changed from whether it is ethically justified to do military (Hersh, 2001) research in support of the defence of one's own country to whether it is ethically justified to do research which may result in weapons which are:
- Traded to countries with poor human rights records and may be used in internal repression.
- Used offensively in support of the country's political and economic aims or not used, despite heavy resource consumption.

The availability of military funding and the fact that many military projects are technically challenging increases the pressures and temptation to become involved in them. For instance over half of US scientists and engineers had military contracts and 65% of US federal research money went into military or related projects in 1990 (Crowe, 1990). The very existence of modern military forces is totally dependent on the participation of scientists and engineers in supporting the military, including through research and development. The existence of military research and the relatively high resources put into it contribute to the build-up of arms races and the development of an ethos which accepts war and preparation for it rather than peace.

Expenditure on military research also diverts resources from social programmes, including education and the health service, and non-military defence strategies (Evans et al, 1991). Resources put into military research and development starve the civil sector of expertise which may already be in short supply (Gummett, 1986) and could be used to solve environmental, social and economic problems, including that of food shortages. At least two studies have found an inverse relationship between the share of GDP spent on military research and development and the rate of productivity growth and international competitiveness in manufacturing (Winn, 1984; Kaldor et al., 1986), though the authors of the second study are cautious about its interpretation (Evans et al., 1991). Military research is also responsible for the developments in armaments technology which have contributed to spreading the effects of conflict more widely, including by increasing the range of many weapons, the speed at which military personnel can be moved and the extent and severity of short and long term environmental damage. Although it has been suggested that military research could be used to reduce the destructiveness of war by developing highly accurate weapons, in practice technological developments have had the opposite effect of blanket bombing of whole areas and increasing the proportion of civilian casualties.

In addition to obviously military research, there are a number of 'fuzzy' areas, including
- Apparently civilian research projects with military funding
- Unforeseen military applications of research
- Military projects relating purely to damage reduction, including the detection of nuclear and chemical weapons.
- Work with both civilian and military applications.

Many apparently civilian areas of research also have military applications, not all of which can be foreseen in advance. However totally refraining from research would not avoid ethical problems, as researchers could be considered to have some ethical responsibility to make their skills available for the benefit of society, particularly when their training has been publicly funded. Work funded by the military raises a number of ethical questions, as there are generally restrictions on the right to publish or permission is required. As well as impeding the free exchange of information, this could lead to the suppression of certain results or small numbers of researchers monopolising funding in particular areas, giving them an unfair advantage over colleagues (Shrader-Frechette, 1994). Carrying out civilian research in civilian institutions and with civilian

funding is more likely to result in a higher degree of public accountability, whereas military funded research may be classified (Lappé, 1990) and is more easily coopted by the military. In addition military establishments generally fund work, even in apparently civilian areas, because they expect it to have military applications. In some cases, as with the UK Ministry of Defence, military applications are disguised by projects being split among different universities (Evans et al., 1991). The control of a high percentage of research funding by the military (and industry) has probably already resulted in a shift away from long term fundamental research to short term applied work with immediate applications, resulting in a de facto limitation of academic freedom and a possible move away from the areas of greatest social or academic merit. However the choice may be between accepting military funding with all its potential problems or not carrying out a particular piece of research.

3. ANALYSIS OF MILITARY RESEARCH WITH MILITARY FUNDING

The various ethical theories will first be applied to the relatively simpler case in which both the main applications of the proposed research and the funding sources are military. In terms of deolontological ethics, many researchers genuinely believe that military research is necessary to ensure that the researcher's country is able to defend itself adequately and to prevent worse evils. Although often based on a rather negative view of human nature and a limited approach to national defence which does not conceive of more peaceful approaches, researchers with this type of motivation can be considered to be ethically justified in terms of deolontological ethics. Other motivations, such as the excitement of solving interesting technical problems and the availability of funding can be considered ethically neutral in terms of deolontological ethics. However, this illustrates the inadequacy of considering purely the ethics of the motivation and intention, without consideration of the consequences.

In terms of positive utilarianism military research can be considered to provide benefits to the participating researchers and possibly their country by acting as a deterrent or increasing its ability to defend itself. There may also be some benefits with regards to so-called spin-offs i.e. civilian applications of the research, though applying the same resources directly to civilian problems would probably be more effective. The risks and costs are the opportunity costs of the resources used in military research and the serious risk of increased casualties and destruction resulting from more 'effective' weapons, as well as the risk of increased preparedness for war significantly increasing the likelihood of war taking place. Thus positive utilarianism does not give an unambiguous view of the ethics of military research and the conclusions drawn from it depend on how the risks of increased casualties and likelihood of war are weighed against the benefits of a more effective deterrent or defence and possible spin-off applications. With regards to negative utilarianism the potential present and future harms of increasing the likelihood of war by increased preparedness for it, the likelihood of increased casualties and the opportunity costs of foregone research or other activities are weighted more highly against the potential benefits. Thus military research would generally be considered unethical in terms of negative utilarianism. This analysis also shows the importance of combining deolontological and consequentialist, such as utilarian approaches.

In terms of virtue ethics, carrying out military research may require some self-deception as to its likely consequences, as well as deliberately or unintentionally ignoring peaceful approaches to conflict resolution. However this may be combined with a genuine love of country and wish to protect it, as well as possibly rather less positive feelings about other countries. Thus there may be a mixture of positive and negative effects on the characters of researchers with this type of motivation. Those motivated by the opportunity to solve interesting technical problems are totally ignoring wider ethical and other considerations. The effects on character of those motivated largely by the availability of funding are likely to be even more negative. Therefore in general, military research should be considered to be unjustified in terms of virtue ethics, though there are exceptions.

Basic moral rights should be considered to include the rights to live in peace and have access to an equitable share of the world's resources and employment, as well as to self-defence which takes into account the rights of other individuals and nations. Military research generally occurs in a world climate which prefers military preparedness to the peaceful resolution of conflict. It contributes to a world situation in which there is preparation for war rather than peace and may reduce the efforts put into peaceful approaches to resolving conflict. Since military hardware is generally very expensive and the context in which military research takes place also resource intensive, supporting military research will generally reduce rather than increase employment opportunities. Although military research contributes to the ability of some countries to defend themselves, it increases existing inequalities between countries in their abilities to defend themselves. Therefore military research will have a tendency to increase rather than reduce existing inequalities between individuals and between nations. Thus military research should be considered unethical in terms of rights ethics.

Military research frequently results from a worldview which can be considered to run totally counter to

ideas of preserving relationships or developing communities, whereas there are alternative approaches to defence based on developing relationships and building trading, cultural and other links with potential enemies to make war less likely. The development of weapons, particularly modern weapons, often takes place in a context which ignores the humanity of those the weapons will be used against. Therefore military research is unethical in terms of the ethics of care.

Overall military research is likely to reduce autonomy due to the great disparity in resources between different nations. This disparity leads to a reduced ability of smaller or poorer nations to defend themselves against larger, richer or more powerful ones. On the other hand military research increases the autonomy of nations which have the resources to carry out high level military research, but such nations generally already have a high degree of autonomy, though it does not follow that their citizens do. The exception to this may be the nations which have a high level of development of technology, but which are not large or otherwise powerful. However this increase in autonomy relative to other nations is generally at the expense of a diversion of resources from social services and may lead to high levels of militarisation and is therefore generally not beneficient. It is probably in the interests of justice that all nations should be able to defend themselves, but counter to it that some nations have a much greater ability to defend themselves than others. Therefore military research generally does not promote justice, beneficience and autonomy and runs counter to normative ethics.

Military research contributes to developing a type of society with an ethos based largely on force, competition and deterrence, rather than cooperation, collaboration and the development of mutual relations of trust. With regards to the ethics of experimentation there has been little effort by governments and other bodies that fund military research to inform people of alternative social models so they can make an informed choice. Individual researchers can make decisions as to whether or not they participate in military research, but few citizens have the opportunity to make informed or indeed any choices as to whether they wish the society in which they live to engage in military research. Thus military research is unethical in terms of the ethics of experimentation.

Military research is frequently very specialised and generally ignores the types of interconnections and wider world views which are often at the basis of research on peaceful approaches to conflict resolution and building a peaceful society. Research into increasing the accuracy and precision of weapons seems to have been noticeably unsuccessful in practice. However, if successful, it could have a significant effect in reducing the negative environmental consequences of war. Any such potential effects are outweighed by the increased preparedness for war resulting from military research and the increases in the potential for social and environmental catastrophe resulting from (recent) developments in chemical, biological and nuclear weapons, as well as the significant but lower levels of potential damage from depleted uranium weapons, cluster bombs and landmines. Thus military research, as currently carried out, is unethical according to ecocentred ethics.

It is the public interest component of professional codes that needs to be considered in evaluating military research in terms of rule based ethics. It will be assumed here that the public interest should be considered to be increasing equity, distributing resources (more) fairly and supporting human and civil rights. As discussed in this and the previous section military research does not contribute to any of these goals and should therefore be considered unethical in terms of rule based ethics. However, in terms of specific professional codes, there may be conflicts between responsibilities to employers and to the wider public.

4. ANALYSIS OF 'CIVIL' RESEARCH WITH MILITARY FUNDING

In this section the different ethical theories will be applied to the 'fuzzy area' of civil projects with military funding. Due to the limited availability of funding, many researchers may consider that they do not have any choice about accepting military funding and some may even consider that they are diverting resources away from truly military projects. Other researchers may consider the research so important that the nature of the funding source fades into insignificance in comparison with ensuring it is carried out. However, researchers may also be influenced by pressures to accept funding or the increase in prestige and opportunities that often accompany it. Thus deolontological ethics gives ambiguous conclusions about the ethics of military research. While not implying that it is highly ethical, it is not necessarily unethical either.

As discussed in section 2, military funded research is likely to have military applications, even if they are not immediately apparent, there may be restrictions on publication and resources will be diverted out of certain areas of research, including long term fundamental research and into others. This could have potentially very serious consequences for the future of research. On the other hand it is possible that some areas of research will not be funded if military funding is not accepted, but likely that (related) research will take place, though probably with less funding, after delays, over a longer period of time and with different researchers. Acceptance of military funding can also reduce the pressures on

governments to provide more civilian sources of research funding. Thus the present and future harms resulting from accepting military funding for civilian research are potentially serious and this should therefore be considered unethical in terms of negative utilarianism. However the situation with regards to positive utilarianism is more ambiguous and depends on the importance given to the particular research being funded (in the short term) compared to the risks and costs discussed.

With regards to virtue ethics, accepting military funding rather than looking for alternative sources of funding or carrying out work without funding may be the easy option. Therefore, in general it is unethical in terms of virtue ethics. However there may be exceptions where researchers carry out intense soul searching and examination of conscience and consequences. In terms of rights ethics, acceptance of military funding may act against the right to have research funded from civilian sources.

With regards to the ethics of care, there could be some tension between researchers who do and do not accept military funding for civilian work. Restrictions of publication and dissemination of information associated with military funding may also impede or prevent the development of communities of researchers. Therefore accepting military funding is unethical according to the ethics of care. Acceptance of military funding by some researchers may reduce the autonomy of other researchers by incrementally contributing to the acceptability of this source of funding and reducing the total civilian funding available for research. Although the results of the research may benefit other people, the acceptance of military funding in itself is not a beneficient act and may impede the funding of other areas of importance in terms of either solving important social problems or developing fundamental knowledge. There is also a certain amount of injustice involved in some researchers accepting military funding, both as it reduces the total amount of funding available and as military funding is not available to all researchers, whether because their research is of no interest to the military or they are prevented by conscience from accepting it. Therefore accepting military funding is not ethical with regards to normative ethics.

Although acceptance of military funding by individual researchers probably does not have direct and immediate social and environmental consequences, it has an incremental effect on militarising the context in which research is carried out. This is likely to have consequences for society as a whole, which are not publicised, making informed consent impossible. Therefore in terms of the ethics of experimentation accepting military funding for civilian research is probably unethical. In terms of ecocentred ethics, accepting military funding for civilian work is unethical, due to the narrow focus and the general ignorance of the wider context and insufficient concern for possible future consequences arising from the nature of the funding and the applications of the research.

With regards to rule based ethics, the previous analysis implies that it is in the public interest for civilian work to be funded by civilian sources. Whether or not it is in the public interest for a particular project to receive military funding depends on a number of factors, including the importance of the research in terms of developing knowledge or solving social and environmental problems and the likely impacts on the research being delayed or not taking place at all if military funding is not accepted.

5. CONCLUSIONS

The paper has applied the main ethical theories to analyse both military research and civilian research with military funding. This analysis has clarified some of the ethical dilemmas associated with military work, as well as showing their complexity. It has highlighted the insufficiency of using deolontological or consequentialist approaches on their own. The analysis has also shown that, though valid conclusions can be drawn from many of the ethical theories, in isolation each of them presents a rather limited perspective which focuses on only one aspect of the problem.

The approach taken here of applying a number of the different theories to give a full picture of the ethical dimensions of a particular problem has potentially very wide applications. In any given problem some ethical dimensions or theories will be more relevant or have more weight than others and it is these dimensions that will influence decision making. The analysis here has considered the question of whether or not a particular action is ethical. In many cases activities can be implemented in more or less ethical ways and the relevant question is how to avoid or minimise unethical consequences or factors.

This analysis has drawn on the available information about the consequences of military research and nature of modern warfare. However in some cases the 'facts' are ambiguous and open to more than one interpretation. In such cases objective assessment is not possible and researchers will interpret the available data in the light of their knowledge, experience, expertise and, generally, also political perspectives. While not necessarily a problem, it is important that this subjective element in ethical analysis is noted, particularly, as illustrated by the discussion, it is generally necessary to make value judgements about the relative significance of different consequences or factors. These value judgements are likely to be strongly influenced by subjective perspectives.

REFERENCES

Agenda, (1992). *Agenda for Action in the 21st Century*, Produced by the Rio Summit.

Babcock D. L. (1991). *Managing Engineering and Technology*, Prentice-Hall.

Barbour (1995). *Ethics in an Age of Technology: the Gifford Lectures*, **2**, San Francisco, Harper.

Bennet, F. L. (1996). *The Management of Engineering*, John Wiley and Sons.

Callicott, J.B. (1992). In: *Encyclopaedia of Ethics*, 1, L.C. Becker (ed.), Garland, 313-314.

Crowe, L. (1990). The federal government, university science and the social contract. *From the Center* (University of Colorado), **9**(1), 1-2.

Evans, R., N. Butler and E. Gonçalves (1991). *The Campus Connection, Military Research on Campus*, Student CND.

Gilligan, C. (1982). *In a Different Voice*, Cambridge, Mass. Harvard University Press.

Gluck, S. E. (1986). Ethical engineering. In *Handbook of Engineering Management*, J.E. Ullmann (ed.) New York, Wiley, 176.

Gummett, P. (1986). What price military research?, *New Scientist*, 19 June, 60-63.

Hartley, K. and N. Hooper (1995). *A study of the Value of the Defence Industry to the UK economy*, Centre for Defence Econ., Univ. York.

Hersh, M.A. (2001). The Ethics of Military Work, In: *Conflict Management and Resolution*, Elsevier

Hersh, M.A. (2002a). Whistleblowers – heroes or traitors?, *SWIIS '02*, Vienna.

Hersh, M.A. (2002b). Ethical Analysis Of Automation, *SWIIS '02* Vienna.

Kaldor, M., M. Sharp and M. Walker (1986). *Industrial Competitiveness and Britain's Defense*, Lloyds Bank Review.

Lappé, M. (1990). Ethics in biological warfare research. In: *Preventing a Biological Arms Race*, S. Wright (ed.), 77-99, MIT Press.

Lappé, M. and B. Bailey (1999). *Against the Grain*, Earthscan.

Madu, C. N. (1996). *Managing Green Technologies for Global Competitiveness*, Quorum Books.

Martin, M.W. and R. Schinzinger (1996). *Ethics in Engineering* (3rd ed.), McGraw-Hill.

Oldenquist, G. and E.E. Slowter (1979). Proposed: a single code of ethics for all engineers, *Professional Engineer*, **49**(5), 8-11.

Park, P.D. (1995). The ethical engineer and the law, *Eng. Sci. and Education J.*, **4**, 277-289.

Renner, M. (1993). *Critical Juncture: The Future of Peacekeeping*, Worldwatch Paper 114.

Scheffler, S. (1994), *The Rejection of Consequentalism*. Clarendon Press Oxford.

Shrader-Frechette, K. (1994). *Ethics of Scientific Research*, Rowman and Littlefield Publishers Inc.

Vesilund, P. A. (1988). Rules, ethics and morals in engineering education, *Eng. Ed.*, **78**, 289-92.

Winn, J. (1984). The university and the strategic defense initiative, *NEA Higher Ed. J.*, 1(1).

ELSEVIER

IFAC
PUBLICATIONS
www.elsevier.com/locate/ifac

METHODOLOGICAL MYOPIA:
A SURVEY OF TECHNOLOGY DEPLOYMENT ISSUES IN LESS DEVELOPED COUNTRIES

B. Lyng and L. Stapleton

ISOL Research group, Waterford Institute of Technology

Abstract: This paper briefly reviews the contribution of engineering methodology to a view of development which is highly questionable. These methodological difficulties are most especially seen in the information technology development arena where current paradigms continue to adhere to a myopic view of technology as it relates to its cultural context. A number of sources of AT and ICT cultural myopia are reviewed in the literature, but few authors have surveyed them in the context of engineering design and development. This paper provides a survey of the key areas in which engineering design and development method and practice are likely to encounter, or foster, problems. The paper proposes a research trajectory which addresses the myopia of current methodology and suggests fruitful directions in which such research might move. *Copyright © 2003 IFAC*

Keywords: Methodology, systems engineering, social context, developing world.

1. INTRODUCTION AND BACKGROUND

Many western economies have moved gradually from industrial to the, so-called, 'post industrial' phase of development during the latter part of the last century. Smaller states such as Ireland, Singapore and Israel formed part of a "second wave" of countries who managed to jump from being primarily an agrarian/partially industrialised to a post industrial economy based on knowledge capital and information intensive employment (Trauth 2002). It is hoped by policy makers that the developing countries of Africa and Asia can make similar leaps to a post-industrial economy.

2. MODERNISATION AND GLOBALISATION

The Modernisation Paradigm currently comprises the transfer of advanced technology such as ICT's to developing countries leading to economic growth by transposing supposedly successful Western models (Nulens (2001). But it is not always successful.

The principle of globalisation as first identified by Levitt (1983) has led IS and software developers to conclude, or at least aspire to the notion that ICT development, and the development of other advanced technologies, is a universally homogenous practice. Consequently, localisation efforts concentrate on the superficial facets of interface design, and sometimes language. Whilst some argue that perceived cultural barriers are diminishing at a global level, it remains the case that many western and non-western countries have deep rooted cultures which have been embedded over centuries and actively resist the imposition, as they see it, of non-native culture (Walsham 2001). For example, Anglophone, westernised popular culture is being resisted in France, where laws have been passed at national level to protect the French language. It has become self-evident that these barriers cannot be addressed by superficial localisation approaches.

This notion of universality which is embedded in many engineering methodologies is conceptually problematic as it fails to recognise that technology develops in a context sensitive, cultural setting (Kaye & Little, 2002). It also can often embody the corporate and national culture of the IS designer (Trauth 2002). In addition, technology and its subsets (such as information systems development method (ISD)), are considered to be primarily a technical affair, with little or no attention paid to the social impact of the implementation, or the cultural requirements for successful operation (Stapleton (2001)). Therefore the design and implementation of information systems and advanced automation technologies in the west are loaded (intentionally or otherwise) with context sensitive attributes. They represent a particular world-view embodied in the technology itself, a 'techno-culture' (Ihde (1998)). Mismatches between the cultural attributes embedded in advanced technologies and the cultural context of implementation can create serious tensions and, ultimately, jeopardise the successful implementation of the new technology.

2. TECHNOLOGY FAILURE & MYOPIA

The probability of systems implementation failure is also exacerbated when rationally problematic and culturally myopic designs are exported to developing countries with only lip-service paid to localisation. A variety of issues associated with localisation have been shown to contribute to difficulties in this space. These issues can be summarised as follows:
1. Skilled Personnel
2. Local Politics
3. Technical Limitations
4. Rationality Assumptions
5. The Progress Imperative - Technological advance as it's own goal
6. Infrastructure
7. We shall now examine each of these in turn.

2.1 Skilled Personnel

Throughout the literature (Mennecke & West 2002, Heeks 1999, Heeks 2002) low literacy levels and by extension, a lack of skilled users and developers is cited as one of the main inhibitors to ICT and AT success in less developed countries (LDCs). Illiteracy rates in Sub-Saharan Africa are currently running at 30% among adult males and 45% among adult females (World Bank 2002). The lack of capacity to develop and acquire technology, and a low technological consciousness among managers (Nulens 2001), correlate with a shortage of educational standards, poorly qualified trainers, inappropriate government policy, a "brain drain" of the best qualified, and poor remuneration. For example, these have all been cited as major problems in Zambian efforts to improve skill levels among students of computing (Corr 1994).

2.2 Local Politics

Technology in LDCs is often considered to be of symbolic importance, a source of information, and consequently a source of control and power. Thus technology becomes a status symbol for some and a tool of oppression for others (Heeks 2002). Danziger et al (1982) suggested that those who hold power over IT may be perceived as technically sophisticated, thereby providing motivation to take political (and consequently irrational) action in ICT and AT projects.

2.3 Technical Limitations

Whilst various private and public sector initiatives have sought to improve telecommunications in LDC's (Richards 2001) the fact remains that only 3% of Sub-Saharan Africans have access to a telephone line or a mobile phone (World Bank 2002a). Therefore communication technologies upon which so many Western networks are reliant, remain widely unavailable in LDCs.

2.4 Rationality Assumptions

Technology developers to often assume that the western view of 'rational' decision-making is universally appropriate (Grenham (1997), Hirschheim & Newman (1991)). These assumptions are embedded in most technology projects through the systems engineering methodologies and deployment approaches, but are unlikely to be the sole legitimate basis for planning and management in LDCs. The developers of a GIS system in India realised the importance of religion and mysticism on decision-making. The result was difficulties in the technologies implementation (Walsham 2000). Deep-rooted religious belief systems encouraged strict demarcation and little or no communication between organisational departments. This suggests the need for engineering methodologies, which are either rationality independent (based on, for example, a meta-rationality), or highly localised.

2.5 The Progress Imperative

The western concept of advanced technology is that these technology implementations are a means to an end. Writers argue that successful projects result in rationalisation, cost saving, improved customer satisfaction, etc. Contrary to this is the LDC perception that technology is a source of coercive power. Bhatnagar (1992) found that African governments considered IT as a goal in itself and not as a development tool. Such technology is a value-laden entity and as a result a status symbol for those who control the technology, and sometimes a tool of oppression (Heeks 2002). Indeed, this view is a reflection of a deeper help assumption concerning economic growth. In this view 'progress' is seen to have enriched the few at the expense of the many. Indeed, one writer argues

'innovations must only be permitted when it is clear that society and the environment will benefit' (Douthwaite (1992) p. vi).

Douthwaite (1992) argues that capitalism must have progress in order to survive. Continuing technological development and innovation contributes significantly to this imperative and ensures that capital is more and more concentrated in the hands of the few. This concentration of economic power is likely to be to the detriment of those who are currently outside the power-axis.

2.6 Infrastructure

*The fundam*ental supporting infrastructure for the correct implementation and maintenance of as IS are often assumed to exist in the local context. Examples include stable electricity supply, certain temperature ranges, ease of acquisition of consumables, and (as alluded to earlier) qualified technical staff, familiarity with a QWERTY keyboard, familiarity with a GUI interface, and even a knowledge of English are often assumptions embedded in systems engineering design methodologies (Westerup 2001).

3. CONCLUSION

The design and deployment of advanced technology has built into it perspectives which may be entirely inappropriate in non-westernised settings. This is exacerbated by the export of the designs to an alien context, without due consideration of customisation of design and implementation challenges unique to that context. To circumvent this problem, systems engineering methods and practices must become more malleable, and thus adaptable to local context. However, as Walsham (2000) put it, these issues can only be addressed "through a detailed and continuous process of learning by actors involved in any particular location." This paper proposes a new trajectory for research in engineering methodology. This trajectory must address localisation of technology and deployment processes into the particular context in which it is to be delivered. This requires new synergies between social impact and deployment management activities, learning solutions and engineering design and delivery. At a deeper level, the debate concerning the nature of technological progress and innovation must be intensified within groups such as IFAC, IEEE, ISD and ACM. Not only is this research investigation

long overdue, but it is critical in the context of increasing international stability in an increasingly polarised world.

REFERENCES

Bhatnagar, S.C. (1992), 'Information Technology and Socio-economic Development' in *Social Implications of Computers in Developing Countries,* eds. S.C. Bhatnagar & M. Odedra McGraw-Hill, New Delhi, pp. 1-9.

Corr, P.H. (1994). 'Computer Studies in Least Developed Countries: A Case Study in Zambia', *Science, Technology and Development,* **vol. 12, nos. 2&3,** pp 270 -280

Danziger, J.N, Dutton, W., Kling, R., and Kraemer, K. (1982). *Computers and politics: High Technology in American Local Government.* Columbia University Press, New York.

Douthwaite, R. (1992). *The Growth Illusion*, Lilliput Press.

Grenham, G. (1997). Rationalistic and Emergent Perspectives on Information Systems Development for Managers, Unpublished PhD. Thesis University College, Cork.

Heeks, R. (1999) ' Information and Communication Technologies, Poverty and Development ' [Online], Institute for Development Policy and Management, Available from http://idpm.man.ac.uk/wp/di/di_wp05.htm [10 February 2003]

Heeks, R. (2002) ' Failure, Success and Improvisation of Information Systems Projects in Developing Countries' [Online], Institute for Development Policy and Management, Available from http://idpm.man.ac.uk/wp/di/di_wp11.htm [10 February 2003]

Hirschheim, R.A. & Newman, M. (1991). 'Symbolism and Information Systems Development: Myth, Metaphor and Magic', Information Systems Research, 2/1.

Ihde, D. (1998). Expanding Hermeneutics: Visualism in Science, Northwestern University Press: Ill..

Kaye, G.R. & Little, S. (2002). 'Dysfunctional Development Pathways of Information and Communication Technologies: Cultural Conflicts' in *Information Technology Management in Developing Countries* , ed M. Dadashzadeh, IRM Press, pp. 186-203.

Levitt, T. (1983) 'The Globalization of Markets, Harvard Business Review (May-June 1983) **vol. 61, no. 3**, pp.92-102

Mennecke, B. & West, L. (2002). 'Geographic Information Systems in Developing Countries: Issues in Data Collection, Implementation and Management' in *Information Technology Management in Developing Countries* , ed M. Dadashzadeh, IRM Press, pp. 71-81.

Nulens, G. (2001) 'Information Technology in Africa: The Policy of the World Bank' in *Information Technology in Context - Studies from the perspective of developing countries*, eds C. Avgerou & G. Walsham, Ashgate Publishing, Aldershot, pp 264-276.

Richards, B (2001) 'The Chicken, The Egg, and African Telecommunications' in *Proceedings of International Conference on Information Technology, Communications and Development*, [Online], Available from: http://www.itcd.net/itcd-2001/papers/doc_pdf/doc_04.PDF [10 February 03].

Stapleton, L. (2001). 'Information Systems Development: An Empirical Study of Irish Manufacturing Companies', Ph. D. Thesis, Department of Business Information systems, University College, Cork, Republic of Ireland.

Trauth, E.M. (2002). 'Leapfrogging an IT Labour Force: Multinational and Indigenous Perspectives' in *Information Technology Management in Developing Countries*, ed M. Dadashzadeh, IRM Press, pp. 221-243.

Walsham, G. (2001) 'IT, Globalisation and Cultural Diversity' in *Information Technology in Context - Studies from the perspective of developing countries*, eds C. Avgerou & G. Walsham, Ashgate Publishing, Aldershot, pp 291-303.

Westerup, C. (2001) 'What's in Information Technology? Issues in Deploying IS in Organisations and Developing Countries' in *Information Technology in Context - Studies from the perspective of developing countries*, eds C. Avgerou & G. Walsham, Ashgate Publishing, Aldershot, pp 96-110.

World Bank (2002). *WDI Data Query*, [Online], World Bank, Available from: http://devdata.worldbank.org/data-query [29 November 2002].

World Bank (2002a). *Key indicators: regional data from the WDI database* [Online], World Bank, Available from: http://www.worldbank.org/data/databytopic/reg_wdi.pdf [29 November 2002].

ELSEVIER

IFAC
PUBLICATIONS
www.elsevier.com/locate/ifac

TECHNOLOGY TRANSFER TO DEVELOPING COUNTRIES AND TECHNOLOGICAL DEVELOPMENT FOR SOCIAL STABILITY

G. M. Dimirovski[1 & 2] and A.Talha Dinibutun[1]

[1] *Dogus University, Faculty of Engineering*
Department of ComputerEngineering, Acibadem, Zeamet Sk. No:21
TR-34722 Kadikoy – Istanbul, Republic of Turkey
FAX#: ++ 90-216 / 327 96 31; E-mail: talhad@dogus.edu.tr

[2] *SS Cyril and Methodius University, Faculty of Electrical Engineering*
Inst. of Automation & Systems Eng., MK-1000 Skopje, Rep. of Macedonia
E-mail: gdimirovski@dogus.edu.tr

Abstract: This is the scond part of a casy study reports on an attempt to develope a framework and conceptual model of optimising development decisions in technology transfer for social stability and sustainable society development. It is grounded on Mansour's systems science approach to the contemporary society systems and of his philosophy of fairness and justice on the presumptions laid dawn by of ethics and ecumenical religious tolerance, and on Dimirovski's strucatural system model of of the interactive impact of human-centred systems in knowledge and technology transfer for development and co-operation. It is claimed that social consciousness world-wide identifies different standards and preconditions that countries of so-called West introduced insofar into tackling the issues of the globalisation have induced individual and collective anti-globalisation opposition processes, regardless the fact that process of economic and cultural globalization is a natural developmental phenomenon within contemporary societies of Mankind. The issues and problems of a fairness transfer of both knowledge and technology to less and developing countries may well be a crucial tool for enhancing the positive effects of the globalisation. *Copyright © 2003 IFAC*

Keywords: Conflicting interests, sustainable development, knowledge transfer, social stability, technology transfer.

1. INTRODUCTION

Mankind of our world has faced a certain rapidly growing and expanding opposing civic reaction while entering the 21st century and as the actual globalisation process as well as political and scientific discussion forums are expanding and quickening. Wherever worl-wide political forums took place, everywhere wide anti-globalisation protest with international participation took place, and the last one in Italy turn out in acivic drama. This opposing civic reaction is not only internationalised, but getting more and more organised so that roomers about a new Anti-Globalisation International are already in the air in many countries worldwide. Yet, it can be argued that

39

the process of economic and other forms of worldwide integration is a historical stage of the development of Mankind on this globe, hence it is inevitably taking place.

Why such an internationalised and growing opposing civic reaction is taking place?
Under the auspices of IFAC, the worldwide community of systems scientists has social responsibility to search and find some part of the answer to this question, or rather a set of answers to issues involved. It may be argued with good reasons, a search to find a proper understanding of the globalisation process and justifiable means and ways how to enhance the positive effects and suppress the negative ones (e.g., see Cuenod and Kahne, 1973; Kopacek, 1995; Hancke and Craig, 2000; Dumitrache, 1997; Dimirovski, 2000).

2. PHENOMENALISTIC MODDELING OF HUMAN CENTRED SYSTEMS, SOCIETY AND SUSTAINABLE DEVELOPMENT

Contemporary societies of Mankind, in both micro and macro levels, are human centred systems possessing individual, local and society-wide awareness and consciousness that act similarly to control and management algorithms. Authors belive that Mansour's premise on fairness and justice is fully compatible with phenomenalistic, systems science based view on contemporary societies (Mansour, 2000, 2001), and the essential issue of sustainable development within the current globalisation process. For from the micro to macro level the word is about interacting human-centred systems, and so is the process of sustainable technology transfer for enhancing economy and society development in underdeveloped countries.

Let us recall briefly the system science approach to model identification in phenomenalistic modelling of humena centred systems, which is summarized in (Mansour, 2000) and presented in Figure 1. By and lagre, the word is about iterative process of managing interest-conflicting and/or criteria-conflicting processes via applying feedback based iterations. So, why not the fairness and justice be among the major criteria in re-shaping world-wide system of global development in economy, hence in technology and culture in the broader sense? Afirmative conclusion in this regard is further supported if we heave a closer look into structures of behaviour of individuals, groups and societies as depticted in Figure 2, which due to size is included at the end of this paper. Clearly again the word is about iterative process of managing interest-conflicting and/or criteria-conflicting processes via applying feedback based iterations.

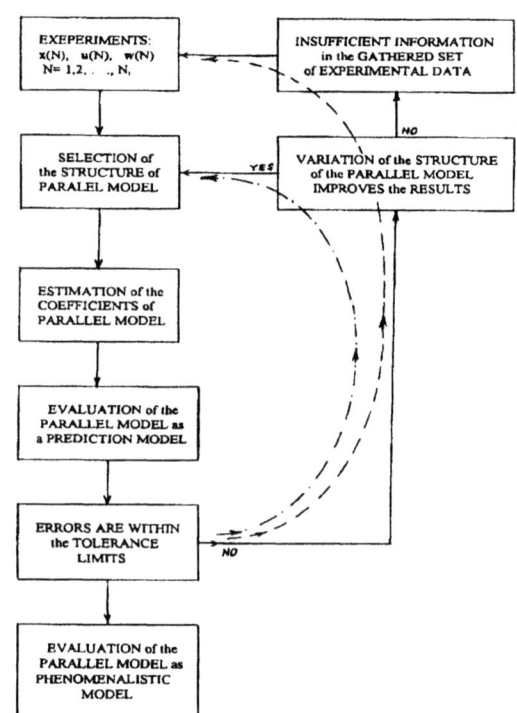

Fig.1. Iterative solving procedure of phnomenalistic system model identification (Mansour, 2000).

2. JOINT KNOWLEDGE AND TECHNOLOGY TRANSFER CAN CONTRIBUTE TO GLOBAL SOCIAL STABILITY DEVELOPMENT

The process of technology transfer is deep into the underlying problems arisen in the current globalisation process and its main controversy implied. It may well for some find it inappropriate to mention it in a research paper, however, the the concerning controversiy cannot be avoided: Is technology transfer just a tool of the new form of colonising and colonialism or a tool of enhancing the economical advancement of the Third World and bridging the North-South gap? Some may well find it naïve, however, authors believe in the heart of the underlying problems arisen lie the issues of fairness and justice and realistically tangible equal opportunities for all, meaning for all countries and nations and not solely for individuals.

In the very essence, it concerns the fairness and justice in feed-back controlled iteravive process of within the globalisation of society development as put forward by Mansour in his plenary lecture at SWIIS 2000, and the actual reality world-wide and in developing countries, in particular, which demonstrates the oposite. The continuing tragedies on the Balkans, Near East and elsewhere (Dinibütün et al., 2000; Richardson, 2001; Scheffran, 2000) all lie in the heart of the globalisation process, also involving technology transfer and re-transfer, and are appealing for more fair and justifiable re-

40

distribution of wealth without which basic social stability and let alone international social stability can be expected (Mansour, 2000).

Let us now point out the systems scince based result on the dynamics of structural changes caused by a serious crisis that dramatically change socities, which is summarized in Figure 3 below. In addition, it should be noted that contemporary world-wide closely interacting and interconnected societies supported by tremendous technology advances in the past century have dramatically shortened the dominant time scale of society development dynamics.

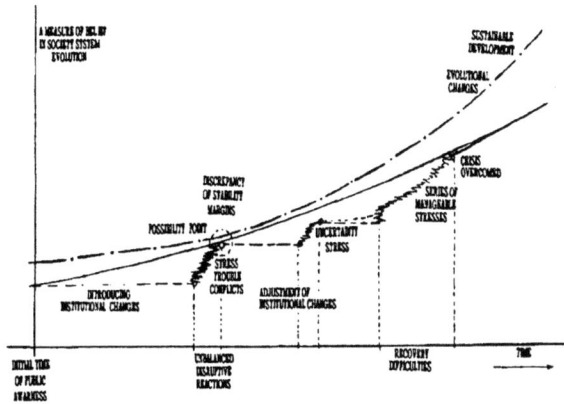

Fig. 2. Developmental dynamics of the process of transitional crisis in Southeast-European countries: more that ten years elapsed and solely Slovenia has reached the state 'crisis overcome' (Dimirovski, 2000).

The real-life experience of the last ten years only has demonstrated that education as well as science and technology do not remain neutral, no matter how and why they have been conceived initially. For, the motivation behind is tightly linked with competition and with tendency for prestigious position within the competition at best, and hence pre-conditioned by given human society and its features. This inevitably enhances conflicting interests, even crisis, unless in its systematic disposal for the benefit of mankind at large on a reasonable time scale is ensured. It is here where ethical categories of fairness and justice can play the role of primary incentives (Mansour, 2000; Dinibütün and Dimirovski, 2001).

The insofar existing practice worldwide, without realy fruitful knowledge transfer, has been producing some more or less short-lived or dead-end intermediate advances (see, for instance, the relevant papers in Hancke and Craig, 2000, or recall the economic failure of most Far-East countries and some Latin-American countries). Therefore these are prone to inefficiency or doomed to failure sooner or later.

It is therefore believed that a directly enhancing and supportive sustainable development approach to developing countries aimed honestly at worldwide integration on equal opportunity basis for all nations precisely by jointly using knowledge and technology transfer among majors tools next to investments is one of the key issues for the globalisation process to be viewed in terms of developments and not a new version of colonialism. The new world order of peace and prosperity on a global scale will seem feasible in such aframework only. And tightly joined technology and knowledge transfer to developing countries seems to be one of the essential processes of control and management to achive it.

How realy important is the knowledge, hence the education, is may well be seen from the recent rather critical reappraisal studies by Frank (2000) and Hanus (2000) concerning the situation in highly developed countries of EU. On the other hand, the joint knowledge and technology transfer trough direct co-operation, and hence education enhancement, has demonstrated its developmental vitality during the last several decades. Of course, this alternative way of handing the problem of concern, among other issues, does require implementation of principles of fairness and justice (Mansour, 2000). To achieve this kind of situation to be a permanent albeit evolving process, the technology transfer into less and underdeveloped countries must be accompanied with adequate sharing of knowledge in order to make technology transfer development enhancing the progress of economy, hence society. Practical implementations may take on various systematic procedures, and some of the recent ones emanating from medium-size Mid-European countries (Kovacz, 2000; Kopacek and Rommens, 2001) and from some other countries (Bloch, 2000; Richardson, 2001) provide guidelines for this purpose.

It is more than evident that all sorts of knowledge in the IFAC fields of expertise, no doubts, play a crucial role. For from the viewpoint of the essential role and impact of technology on the achievements of better efficiency, quality and productivity of industries and services within the competitive world of economy, national and trans-national, indeed these are of paramount importance (Savona and Jean, 1997). However, automated technologies are among the most expensive ones, and so is the knowledge associated, and moreover provide the cutting edge for the economy competitiveness thus not easy amenable to sharing. Therefore the principles of fairness and justice have to be implemented as incentives in both knowledge and technology transfer for the sake of common worldwide benefit on a longer run. Perhaps, we should make it clear, while enhancing the transfer of knowledge and technology that this process is costly in itself, and time consuming due to the

41

sophistication involved. At the same time a considerable serious care must taken that transfer of automation and control technologies be directed solely to civilian sectors and human and social needs, that is the industrialisation (Bloch, 2000).

Authors believe that is rather important from the point of view of IFAC's cosmopolitan role, while taking the relevant care both with regard to the actual state-of-the-art in high-techs and with regard to speeding up a certain sustainable development in developing countries, that care is taken to increase the worldwide awareness on the usefulness and potential efficiency of low-cost automation systems. In particular, should the emphasis is put on obtaining better quality products/services and higher efficiency in using energy and raw materials, the awareness should be increased that all production and service technologies can be upgraded to higher competitiveness from mass application of low-cost control technology (Dinibütün and Dimirovski, 2001).

4. DEVELOPMENT DECISIONS IN KNOWLEDGE AND TECHONOLGY TRANSFER

Systems approaches to developmental problems in general do imply forecasting and planning, but require also relevant decision-making and optimising analysis, and only then management execution. In particular, this is the case with processes of transfer knowledge and technology to underdeveloped countries. It has become apparent that knowledge and technology transfer, which reduced the so-called North-South gaps in economy and social welfare, cannot be properly accomplished on a day-to-day basis or solely following some intuitive decision policies.

It is emphasised, however, that in doing so the way od systems science and engineering we have to confront the challenge of dealing with composite qualitative and quantitative factors as rigorous as feasible in order to achieve optimising development decisions (Gresford, 1972; Kahne, 1973).

The above comprehension of issues discussed insofar has led us to revisit Kahne's work, which offers a systems engineering methodology to tackle joint the knowledge and technology transfer problem on the background framework constructed in this and the previous paper (Dinibütün and Dimirovski, 2001) of ours. Therefore we find it constructive to cite one of original concluding arguments put forward by Kahne long ago, whose work initiated this line of research.

Namely, one can read in Kahne's paper (1973) this argument:

"The problem (i.e., the planning and development decisions making – our emphasis) is separated into five distinct parts. Each part is discussed in the context of the planning process and each leads logically to the optimisation of development decisions. The five parts are goal definition, establishment of criteria, criteria weighting, alternative rating, and alternative ranking. The feature of making development decisions, which distinguished them from other optimisation problems, is what has been called 'fuzziness'. In any realistic problem formulation, the criteria are not precisely defined; they are fuzzy. The relative importance of each criterion is also fuzzy. Indeed, even one attempts to rate a particular possible solution, he must deal with fuzzy information." – S. Kahne

5. CONCLUSION

The globalisation processes, no doubts, shall continue because Mankind's societies and national economies worldwide have come to that stage of development. Therefore both the endeavours and the efforts on concerted activities of the IFAC should converge so as to establish more effective actions focused on enhancing jointly both knowledge and technology transfer to countries around the globe.

The process knowledge and technology transfer does not seem to deliver the expectations if it is left over solely to traditional policy making authorities via copying of existing experiences. System scientists and engineers as well as technology management specialist should enter on a much larger scale into the issues and problems involved. It is their social responsibility to collaborate in elaborating the most adequate models and procedures for effective and efficient conduct of knowledge and technology transfer according to the needs in particular regions and countries.

In this paper, we have argued and given evidence the technology transfer is inseparable from the knowledge transfer both for the reasons of pragmatic needs, and even more so because of the reasons of ethics and social responsibility of individual scientists and engineers, their professional societies and international associations of societies. It seems no new world order guaranteeing peace and sustainable socio-eceonomic, hence political, stability can be achieved without emphasis on the fairness and justice paradigm.

REFERENCES

Bloch, A. (2000). Technologies transfer to a developing country the road to industrialization. In: *Preprints of the 1st IFAC Conference on*

Technology Transfer in Developing Countries: Automation in Infrastructure Creation (G. P. Hancke and I. K. Craig, eds.), Pretoria (South Africa), pp. 161-166. The IFAC and University of Pretoria, Pretoria.

Cuenod, M.A. and S. Kahne, Eds. (1973). *Systems Approaches to Developing Countries,* Algiers (Algeria). The IFAC and the ISA, Pittsburgh, Pennsylvania.

Dimirovski, G. M. (2000), Applied system analysis of transitional crisis of Southeast Europe and national goals formation. In: *Conflict Managemnt and Resolution in Regions of Long Confronted Nations* (G. M. Dimirovski, ed.), pp. 85-93. Pergamon Elsevier Science, Oxford, UK.

Dinibutun, A.T., M. K. Vukobratovic, N. Shivarov, K. Schlacher, and G. M. Dimirovski (2000), Roads to regional conflict resolution via academic and national goals of co-operation. In: *Conflict Managemnt and Resolution in Regions of Long Confronted Nations* (G. M. Dimirovski, Ed.), pp. 119-125. Pergamon Elsevier Science, Oxford, UK.

Dinibutun, A.T. and G. M. Dimirovski (2001), Systems science view on integration of Southest Europe into EU: Technology transfer issues. In: *Automatic Systems for Building the Infrastructure in Developing Countries* (G. M. Dimirovski, ed.), 281-286. Pergamon Elsevier Science, Oxford, UK.

Dumitrache, I., Ed. (1997). *Preprints of the 6th IFAC Conference on SWIIS,* Sinaia (Romania). The IFAC and Politecnica University, Bucharest.

Frank, P.M. (2000). Does the Green Card solve the problem of Germany's lack of engineers? In: *Proceedings of EPAC 2000: The 21st Century Education and Training in Automatic Control* (G. M. Dimirovski, ed.), pp. 77-80. Institute of ASE, Faculty of EE, SS Cyril and Methodius University, Skopje.

Gresford, G. B. (1972). Systems approach for development. *IEEE Transactions on Systems, Man & Cybernetics,* **SMC-2**, 3, pp. 311-318.

Hancke, G. P., and I. K. Craig, Eds. (2000). *Preprints of the 1st IFAC Conference on Technology Transfer in Developing Countries: Automation in Infrastructure Creation,* Pretoria (South Africa). The IFAC and University of Pretoria, Pretoria.

Hanus, R. (2000), From Montaigne to Candide. In: *Proceedings of EPAC 2000: The 21st Century Education and Training in Automatic Control* (G.M. Dimirovski, ed.), pp. 77-80. Institute of ASE, Faculty of EE, SS Cyril and Methodius University, Skopje.

Kahne, S. (1973). A procedure for optimizing development decisions. In: *Systems Approaches to Developing Countries,* (M. A. Cuenod and S. Kahne, eds.), Algiers (Algeria), pp. 385-392. The IFAC and the ISA, Pittsburgh, Pennsylvania.

Kile, F. (1995). A wellness model for of piece and stability. In: *Preprints of the 5th IFAC Conference on SWIIS* (P. Kopacek, ed.), Vienna (Austria), Paper FR-01. The IFAC and IHRT-TUW, Vienna.

Kile, F.O. (2000), The road to active piece (Key Note Lecture). In *Conflict Management and Resolution in Regions of Long Confronted Nations* (G. M. Dimirovski, ed.), pp. 11-16. Pergamon Elsevier Science, Oxford, UK.

Kopacek, P., Ed. (1995). *Preprints of the 5th IFAC Conference on SWIIS,* Vienna (Austria). The IFAC and IHRT-TUW, Vienna.

Kopacek, P. (2000), SWIIS – An important expression of IFAC's commitment to social responsibility (Survey Paper). In *Conflict Management and Resolution in Regions of Long Confronted Nations* (G. M. Dimirovski, ed.), pp. 17-21. Pergamon Elsevier Science, Oxford, UK.

Kopacek, P. and E. Rommens (2001). Control technology transfer: Universities and SMEs – the Austrian approach. In: *Automatic Systems for Building the Infrastructure in Developing Countries* (G. M. Dimirovski, ed.), 271-274. Pergamon Elsevier Science, Oxford, UK.

Kovacz, G. L. (2000). Technology transfer results achieved by Hungarian SMEs and academic institutions. In: *Preprints of the 1st IFAC Conference on Technology Transfer in Developing Countries: Automation in Infrastructure Creation,* Pretoria (South Africa), pp. 150-154. The IFAC and University of Pretoria, Pretoria.

Mansour, M. (2000), Systems theory and human science (Invited Plenary Lecture). In: *Conflict Management and Resolution in Regions of Long Confronted Nations* (G. M. Dimirovski, ed.), pp. 1-10. Pergamon Elsevier Science, Oxford, UK.

Richardson J. (2001), The new petroleum pipeline-terminal project in Chad/Cameroon: Ethics, fairness and justice in the application of technology. In: *Automatic Systems for Building the Infrastructure in Developing Countries* (G. M. Dimirovski, ed.), pp. 275-280. Pergamon Elsevier Science, Oxford, UK.

Savona, P. and C. Jean (1997), Geoeconomica: Il dominio dello spazio economico. A cura di Paolo Savona et Carlo Jean, Franco Angely s.r.l., Milano.

Schffran, J. (2000), Power distribution, coalition formation, and multipolar stability in international systems: the case of Southeast Europe. In: *Conflict Management and Resolution in Regions of Long Confronted Nations* (G. M. Dimirovski, ed.), pp. 37-48. Pergamon Elsevier Science, Oxford, UK.

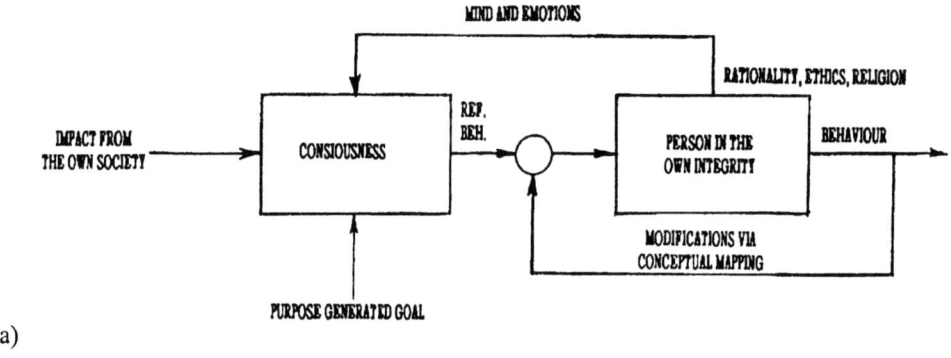

(a)

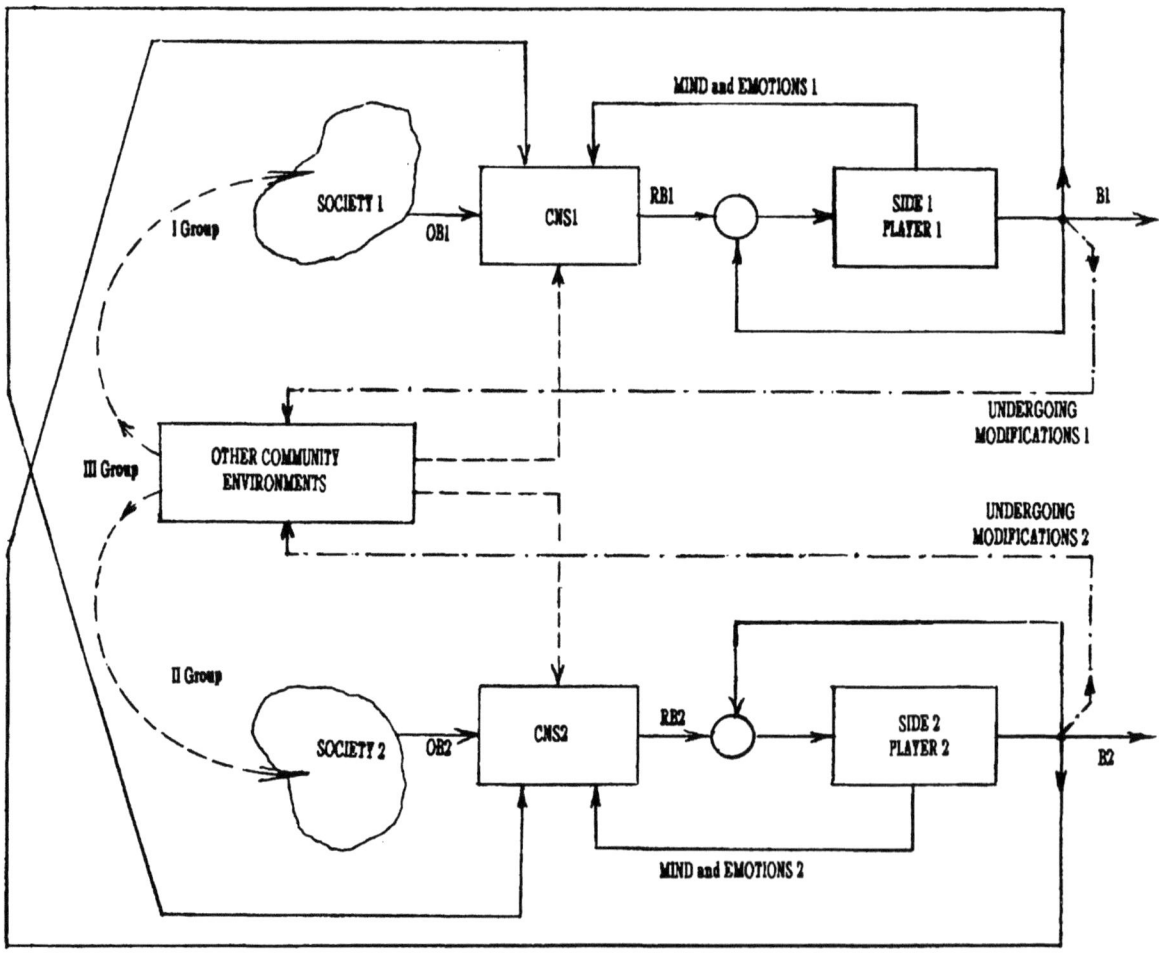

(b)

Fig. 3. System structure in a conflicting situation among contemporary societies: (a) Modified Mansour's (1998) of individual conduct; (b) the underlying overall multi-loop, feedback-feedforward control system architecture of two sides (human centred sub-systems) derived (Dimirovski, 2000).

ADVANCING BEYOND A MULTI-POLAR WORLD

F. Kile

Microtrend
Appleton, WI 54913-7181
U.S.A.

Abstract: Since 1991 the world has moved rapidly from a bipolar political standoff through a period of multi-polarity. This multi-polar world is less well understood than the bi-polar world of the Cold War period. There is reason to believe that this complex world will never be understood from a theoretical standpoint. However, as control engineers and colleagues seek to enhance global stability in this nearly chaotic situation, it is imperative to be aware of resource and ecological factors which may cause some approaches to stability to be unworkable. Limits are related to pollution, resource depletion and exhaustion of the environment's capacity to support society at desired levels. In extreme cases, environmental exhaustion will lead to rising death rates, large-scale migration, and conflict. This paper examines the nature of these limits. *Copyright © 2003 IFAC*

Keywords: Communications, Energy, Environmental, Global, Limits, Stability

1. INTRODUCTION

From 1991 to 2003 the world has moved rapidly from a bipolar political standoff through a period of multi-polarity. The bi-polar world of the Cold War period is fairly well understood. The multi-polar world of the 1990s and early part of the 21st century has been less well understood, in part because it was a far more complex world than the bi-polar phase of history which preceded it. Even as control engineers and others seeking to enhance global stability began applying their skills to tasks helping to stabilize disequilibrium marking the shift from bi-polarity to multi-polarity, parameters affecting global equilibrium (international stability) have begun shifting.

1.1 Evolving international stability parameters

The effects of environmental change are becoming more quantifiable in the early 21st century than these same effects were in the early 1990s. At that point there was relatively strong anecdotal evidence for change affecting large-scale ecosystems, but quantitative data were subject to varying interpretations. By 2003 there were solid quantitative studies of the effects of global warming and related changes in migratory behaviors of many species, notably several species of birds. This paper assumes that aggregate human activity is at least partially a causal factor for global warming. Control engineers with experience in environmental issues may be able to assist in analyzing interactions among large environmental systems. This experience will be helpful in identifying natural systems threatened with collapse absent special intervention.

Environmental change encompasses factors beyond global warming. These changes include altered drainage patterns for river basins, deforestation by migrating human populations, concentrations of pollution in the vicinity of rapidly growing urban centers, and depletion of some species of marine life, among other factors.

1.2 Major premise

A major premise of this paper is that achieving international stability during periods of rapid global environmental change may not be possible. Many aspects of human activity are dependent on relatively predictable growing conditions in major agricultural regions. The influence of climate change is especially amplified in areas which provide subsistence living for local populations during periods of climatic stability but which cannot sustain local populations during periods of climatic instability.

1.3 Secondary premise

A secondary premise of this paper is that the apparently balanced multi-polarity of the past dozen years is changing rapidly. A new phase in international dynamics is emerging. This emerging phase in international dynamics evidences aspects of chaos as much as it evidences a transition from one pattern of international dynamics to another identifiable pattern. A new dynamic pattern may emerge, and it would be helpful to identify a path from a former stability regime to a subsequent regime during the periods of nearly chaotic instability. Engineers who have studied relationships between stability and turbulence and other chaotic behaviors may be able to offer insights into strong and weak stability under emerging international conditions. The SWIIS Technical Committee is an excellent forum in which to discuss relevant insights.

1.4 Current instability is nearly chaotic

During the 1990s it appeared that the UN, in cooperation with the remaining military superpower, could maintain a weakly stable international dynamic through a period of transition in which some nations would develop rapidly. If this had occurred, the developing world would gradually give way to a developed world, albeit some regions would continue to be more highly developed than others. Development along these lines no longer appears to be a likely outcome. What has changed to interrupt the formerly assumed steady progress toward development for most of the world's people?

Population pressure disrupted regions which appeared to be on a path toward development. Food resources became stretched due to population pressure. Agricultural production did not grow as rapidly as population, especially in Central and Southern Africa and also in parts of Central Asia.

Changing patterns of commerce and trade amplified the rapid spread of AIDS, crippling the ability of some areas to cope with high birth rates.
The spread of worldwide communications alerted millions of people to the possibility of migration away from areas of poverty and starvation to more desirable areas. Controls for migration patterns are largely unchanged from an era in which flight from political persecution was the main reason for migration. In the new dynamic, migration for political reasons is relatively stable while flight for economic reasons is growing. Migration has two major effects: it often drains capable people from troubled regions; at the same time, regions receiving large numbers of immigrants are no longer able to offer new populations opportunities open to earlier waves of immigrants.

The spread of communication highlighted real differences in per capita material well-being and created a climate of resentment in areas which were less prosperous or nearly destitute.

Terrorism which had been used for centuries by political leaders seeking change became a weapon of protest against differences between regions.

The ability of governments to meet the expectations of their populations eroded under the multiple stresses of population growth, increasing expectations, mismanagement, and corruption.

Mass communications amplified existing tensions within many national and regional societies. These tensions emerged in the form of radical political groups, radical religious groups, strikes, and impatience of populations for positive change.

The 1990s fostered an illusion of rapidly growing and proliferating prosperity, which seemed both promising and irreversible. Whether or not increasing prosperity can become a reality is not clear. However, it is clear that the spread of prosperity is not a monotonic function. Periods of retrenchment occur. Recent worldwide economic dislocations destroyed many plans for broader participation in the emerging prosperity because planners, both societal and individual, did not allow for dislocations in the pace of growth in prosperity. Many nations and individuals planned and lived from day to day, basing expectations on unrealistic assumptions. As economic growth slowed, bankruptcy and social dislocation spread rapidly. Dislocations have been most severe for groups which felt they were next in line to participate in growing prosperity.

1.5 Environment as framework

The environment is an unavoidable framework for every attempt at international stability. Most participants in debates regarding progress toward international stability discuss environmental factors as though negotiations include the environment. For genuine progress toward sustainable international stability it is imperative for all parties to acknowledge that the environment does not negotiate. The environment appears to have strong stability dynamics. However, when any action, whether human or extrinsic, such as a large meteor impact, moves the environment outside its dynamic range, a new dynamic domain often emerges. Each new dynamic domain seeks its own balance. Scientific knowledge is clearly inadequate to predict characteristics of a new domain before it emerges.

This paper assumes that any human arrangement which purports to enhance international stability and does so in a manner leading to environmental change will not generate stability because human societal activity is heavily influenced by the environment, most notably in the manner in which humans depend on the environment for food, water, and breathable air.

2. CHANGING INTERNATIONAL DYNAMICS

Multipolarity is being displaced by multiple social breakdown. This paper assumes that return to an earlier status quo is not possible. Currently, there is a sharp disjunction between areas which assume that former stability can be restored and areas which assume that progress toward prosperity is an illusion. The latter areas have entered periods of social chaos. A graphic example is the situation in Palestine. The following analysis states what the author believes to be a underlying disjunction in that troubled area.

2.1 Palestine as case study

As a case study, consider the current condition of near chaos in Palestine. The strongest root of dissonance in Palestine is the growing gap between the goals of competing groups. The State of Israel appears to assume that order can be restored and ultimately some accommodation can be reached between the State of Israel and the Palestinian people. This arrangement would constitute a balancing of current demands by each competing group -- a balancing with past realities. The Palestinian group appears to assume that former arrangements can be restored. Neither group recognizes that the dynamic patterns have greatly changed.

The problem in Palestine has grown from a local problem into a regional problem. New arrangements between two parties may not prove sufficient to satisfy regional interests.

The total population of Palestine/Israel has grown substantially since passage of UN Resolution 242 in November, 1967.

Water resources, always scarce in the region, are clearly insufficient to support the entire population within the borders of the State of Israel and Gaza and the West Bank at growing levels of consumption. Available water resources are short of being able both to supply current needs of the State of Israel and a large Palestinian population at current levels of consumption in the State of Israel.

Though both major parties to disputes in Palestine are aware of the discrepancy in water availability and future water demands, this issue in not addressed in a meaningful way in "peace" forums.

The outline of a case study of Palestine, sketched above illustrates an unwillingness of parties to negotiations (or to conflict) to recognize that extrinsic factors may be at least as important to progress toward stability as factors which appear to be intrinsic causes of the initial instability.

2.2 Instability in other regions

The brief sketch of a case study of instability in Palestine can become a template for examining other examples of regional or national instability. Although this paper will not address other possible case studies, we cite a number of instances to alert the reader to why we describe the current international dynamic as chaotic rather than transitional.

At least four nations in Latin America are near bankruptcy.

Food shortages appear each year in impoverished nations. Some shortages follow what appear to be temporary dislocations in weather patterns. Other instances of food shortage or famine appear to stem from the growth of some populations beyond the capacity of their home regions to support the increasing population. In many areas overpopulation harms the local environment and further exacerbates incipient food shortages.

Increasing energy use -- coal, petroleum, and combustion of trees, shrubs and grasses is affecting the global environment in a manner which is partially predictable at best. Global warming, widely thought to be amplified by human use of these combustible fuels, lags actual use by years if not decades. This suggests that, to the extent carbon dioxide amplifies global warming, the warming trend will accelerate.

An ever larger number of developed and partially developed nations are unable to provide meaningful employment to their populations. This is resulting in growing internal strikes and political tensions. Unemployment and its counterpart of underemployment are creating societal dislocations in many areas.

3. TOWARD REDUCING SOCIETAL CHAOS

Three factors appear to be root causes of current

societal disruption. Each of these factors appears to relate directly to the interaction between human society and the natural environment.

Societal negotiations on international issues have largely assumed that the environment is negotiable and that the environment is a party to negotiations. Neither of these assumptions has any basis in fact. Thus, the imagined benefits of the Kyoto agreement on energy use were illusory from the outset, since total global energy consumption would continue to grow, with no foreseeable limit, even though regional and national consumption in many areas would be frozen. The apparent assumption surrounding this agreement was that the environment would respond favorably to the treaty, if the treaty were in force and enforced. The fact is that the global environment will respond only to total worldwide combustion of carbon-based material, whether the combustion is in the form of forest fires, petroleum products, natural gas, coal, or brush, sticks and animal dung. This observation is not a justification for any increase in combustion anywhere. The observation simply acknowledges that the environment responds similarly to a particular perturbation, no matter what the origin of the perturbation.

The second factor causing societal disruption is growing consumption, especially any form of consumption which burdens the environment. One may assume that the environment does not notice or respond to certain forms of consumption. For example, if a person is viewing a television program and a second person joins the first viewer, "consumption" of the entertainment or educational value of the program is doubled with no apparent added environmental loading. However, aside from limited examples such as the one just cited, most consumption has an impact on the environment. In theory, and we believe in fact, every increase of consumption further increases environmental loading. Longer-range effects of this loading are particularly evident in the consumption of water resources, especially in regions of marginal rainfall. In some regions diversion of river flows for new or alternate uses has depleted available water supplies to the extent that significant bodies of water are disappearing completely. Two notable examples are the drastic size reductions of the Aral Sea in Central Asia and Lake Chad in West Central Africa. Both bodies of water, formerly impressive for their size in relativeely dry regions, may vanish within twenty years. Populations which depended on them for sustenance have migrated elsewhere or will do so very soon. Migrations of this type are major social disruptions for the migrating people and for their new homelands.
The third factor causing societal disruption is growing population. Ultimately, each person added to global population increases environmental loading.

The situation outlined in this paper poses many difficult issues. Moreover, examples cited are selected examples. Every informed reader is aware that this outline is incomplete. The most important observations presented here are:

Societal disruption has two underlying causes: population growth and consumption growth.

The environment responds to human activity. It is not a negotiating party to any human agreement. In turn, the environmental response to human activity affects future human activity.

As control engineers, and other persons with influence on societal, organizational, and individual decisions, seek to alleviate social end economic dislocation, we should continue to apply our understanding of the differences between system optimization and sub-optimization.

No group is better equipped to understand the often subtle distinction between systemic optimization and sub-optimization than design and control engineers. We can greatly assist the larger society to deal effectively with current and emerging societal issues to the extent that we apply our unique insights to problems we address.

4. CHALLENGES FOR CONTROL ENGINEERS

48

ELSEVIER
IFAC
PUBLICATIONS
www.elsevier.com/locate/ifac

THE CONCEPT OF UNIVERSAL ETHICS IN GLOBAL NETWORKING – SOME EXAMPLES

C. Rose* and D. Brandt**

**Interdisciplinary Doctoral Programme on Global Challenges
University of Tuebingen
Home Address: Schlossstrasse 80a
D-51429 Bensberg
e-mail: theodortuetchen@gmx.de*

***Dep of Computer Science in Mechanical Engineering (ZLW/IMA)
University of Technology (RWTH)
52056 Aachen, Germany
e-mail: brandt@zlw-ima.rwth-aachen.de*

Abstract: Globalisation under the impact of Information and Communication Technologies requires increasingly the awareness of our own responsibility towards all humans as well as towards the natural environment and towards future generations. Thus the questions of *how to decide* and *how to act* challenge the interdisciplinary research field of *Ethics* more than ever. But how can different societies and social entities dealing with each other be drawn into such responsible behaviour, that means: into agreeing on *one set of ethical rules*? A solution to these questions may only be found through a *universal ethical concept*. As a philosophical solution, K.-O. Apel and J. Habermas (Frankfurt am Main, Germany) developed the concept of *Discourse Ethics*. It works out the basis for joint moral decisions, which are acceptable to everyone, independent from cultural, social or religious background. It claims for all communication partners to be *partners on equal terms*. In this paper, these elements of *communication* and *ethics* are discussed in some detail. Subsequently they are illustrated by selected examples of ethical rules suggested by different organisations. *Copyright © 2003 IFAC*

Keywords: Enterprises, ethics, global networks, technology

1. INTRODUCTION: THE CHALLENGES OF GLOBALISATION

The challenges of globalisation under the impact of information and communication technologies are to be felt everywhere around us. In world-wide settings it has become almost impossible to consider all consequences of our own actions if we do not learn how to think in more general frames of reference. During the last ten years, *globalisation* has emerged as a major catchword in international affairs: in 1990, the word was not to be found in any dictionary at all. In 1993, one of the major German newspapers, Frankfurter Allgemeine

Zeitung (FAZ) only mentioned it 34 times, but from then on its use has enormously increased to the figure of more than 1000 references in 2000.

This development shows how important international co-operation has become in our everyday lives. But even though every person in almost all parts of the world seems to be somehow engaged in the process of internationalisation, there are many more challenges and open questions than solutions and answers can be found by politicians. In his introduction to the Interim Report of the *Enquete Commission on Globalisation*, Chairman Ernst Ulrich von Weizsaecker, Member of the

German Federal Parliament, thus points out that it is an important prerequisite of successful globalisation to motivate and integrate civilian society into the process of internationalisation (Bundestag, 2002).

What are the main characteristics of such global co-operation?

In world-wide settings it has become almost impossible to consider all consequences of our own actions. Not every situation from which we gain short-term win is sustainable with regard to interpersonal or intercultural settings. The questions of how to decide and how to act are the questions of *Ethics*. But not only philosophers have to take up the challenge of answering them. It is the duty of every person him- or herself to consider and discuss processes of everyday decision-finding, be it in engineering, entrepreneurship, science and research, university teaching, or any other activity (Rose et al., 2002).

With *globalisation,* we are discussing the issue of *co-operation.* When we discuss co-operation in economic and technological affairs today, we need to look at how people want to live and work. One of the main aspects, going back to Aristotle and Kant, but now coming up even stronger than ever, is that people want to experience a higher degree of *freedom* in their actions. This claim includes the following two aspects:

Firstly, concerning their individual work, people want to experience *freedom of choice* in planning and organizing their actions. This wish reflects their desire to be accepted by their superiors as responsible human beings who know what they are doing.

Secondly, concerning co-operation and networking, they want to be seen by their partners as *equals* who have freedom of choice in how to develop their co-operation. It then has to be co-operation *on equal terms* if it is to be successful.

This desire for *freedom* and *choice* seems to be an in-bred inheritance of all of us as human beings. But it is, in parallel, one of the central questions of philosophy in all our cultures and civilisations: What does it mean to be free? And, as we act in life and work, what is *right* to do, what is *wrong* to do? How do we decide about right or wrong, positive or negative consequences for all future generations and us? The question of freedom is always the question of what is the *Ethics of our actions.*

2.   THE PHILOSOPHICAL CHALLENGE OF UNIVERSAL ETHICS

As we have seen, globalisation is inseparably connected with the issue of morals and ethics. *Morals* or the *Code of Morals* here means a certain set of rules within a small community, based on traditions and intuition. Ethics means the critical reflection on such sets of rules.

The constitution of any democratic community, for example, is meant to include a catalogue of fundamental rights and rules, which are to be followed in order for all members of the community to respect each other and to live together peacefully. But the content of these rules differs from community to community. These differences tend - in intercultural communication - to produce disagreement about essential decisions. (One example for such difficulties is the international debate on *Human Rights*.) It is a major challenge for all disciplines, but especially for Philosophy, to find answers in this field.

The task of ethics is to reflect and check on moral rules and regulations for human interaction and, by this critical approach, to find *sensible* ways of living together. Behaving according to the *Code of Morals* means voluntarily and consciously to follow certain rules and regulations. It thus concerns our responsibility and our ability of decision-making as human beings. Only living together according to a *sensible* code of ethics can reach beyond the borders of culture or religious faith.

In Western philosophy, ethics as a subject of philosophy goes back as far as to the ancient Greeks, to Aristotle (384 BC-322 BC). Aristotle's considerations include the aspect of virtue or *morals*. He defines this aspect as the opposite of actions which are merely motivated by ignorance or oppression. *Moral behaviour* thus means to follow our own choice, our own measures, and our own reasoning. The questions which derive from Aristotle's approach do, even in times of globalisation, still remain the same: Are we able to make choices? Do we have the options to make our own decisions? We are able to *think* in terms of ethics and to *act* morally only if these conditions are fulfilled.

As a turning point in Western philosophy, Immanuel Kant (1724-1804, Germany - the intellectual driving force in the process of *enlightenment*) subsequently started to consider moral behaviour from the point of view of the *individual*. For him, the only possibility to gain freedom of decision is to *enlighten ourselves*. He challenges us to emerge from the situation of *nonage*. This situation is comparable to the position of children before they grow up, or before they are *coming-of-age*. In this situation of immaturity, children do not have the option to judge about right or wrong. Instead, they are only *following* but not yet *reflecting,* the rules of social action. Only when we dare to make these steps of critical reflection on our reasoning and actions, we can free ourselves from determination by others, e.g. from religious, cultural or political 'oppression'. Kant's appeal of enlightenment, thus, is the Latin 'Sapere aude!': 'Dare to think!'.

In our Western cultural settings, we can still regard Kant's concept of self-enlightening as the most fundamental contribution to a development towards civil rights, freedom, equality, and equal opportunity.

Still, if we follow this tradition, it is also necessary to accept other cultures' philosophical concepts of ethics and discuss their contents *on equal terms* with their representatives. Failure of acceptance concerning, e.g., Eastern, Islamic, Latin American or African philosophies only leads to '*Euro-centeredness*' which seriously affects the efforts of a positive outcome of Globalisation as discussed above. In philosophy, concepts such as *Intercultural Philosophy* respond to these necessities.

3. ENTERPRISE CO-OPERATION OF FUNDAMENTALLY EQUAL PARTNERS

Taking these preliminary considerations into account, one main aspect of *enterprise network co-operation* is equality among regional as well as international network partners. This aim is based on the fundamental recognition that long-term commitment can only develop in co-operation among equals (Rose et al., 2002).

All human beings want to be accepted on equal terms - only then creativity and mutual trust develops. It reflects the age-old desires of humankind: we do not want to work under a tyrant's rule, we do not want to be slaves under a master, we do not want to be remote-controlled by the owner of the enterprise. We want to be free, and to decide for ourselves how we organize our lives and our work.

Only among partners working together *on equal terms,* can mutual trust be established as the most important basis for network co-operation. Only such fundamental trust can create Win-Win situations to the advantage of all partners. But as long as we do not trust each other, we will still feel the need of *competing* instead of *co-operating..* Such competition unfortunately is reinforced by the problems of modern working conditions. Equality and mutual trust have to be established despite the threats of a competitive economic environment, and national and cuy*sal validity* of ethics. In the

following, we will thus try to find an answer to the following question:

How can societies and social entities dealing with each other, be drawn into agreeing on a set of ethical rules, that means: into *responsible behaviour?*

4. DISCOURSE ETHICS IN NETWORKS

The considerations concerning ethics, have, up to this point, shown us the need for free decisions which also refers to co-operation in *networks*. Our aim is to reach fundamental equality of all partners. It means to prevent the network from certain hierarchical structures which endanger freedom of individual participants or minorities.

But even if we achieve this aim, there are always communication problems remaining. Because of time pressure and great distances between ourselves and our partners, there is more and more *virtual communication* taking place instead of face-to-face meetings, and instead of getting to know each others personally. In larger networks, we often have to deal with large or heterogeneous groups, and we might have to face difficulties due to inter-cultural or international and cross-cultural communication problems. Factors like these might lead us to conflicts about ethical or moral concepts and rules, or to misunderstanding each other, and this again can lead to lack of trust. A solution to these questions might only be found by means of a *universal ethical concept.*

This challenge has been widely discussed in Germany as well as internationally. The philosophical concept of *Discourse Ethics* is mainly represented by K.-O. Apel and J. Habermas (both from Frankfurt a. M., Germany, see Apel, 1997). It intends to work out the basis for ethical or moral concepts which are acceptable to everyone, no matter which cultural, social or religious background he or she has. It is obvious that such a concept cannot offer a collection of maxims such as the biblical 'Ten Commandments'. Such rules can only be derived if they are founded on the cultural or religious background of one specific community; and due to this fact, they may contradict those of other communities. Hence, when discussing the *universal* ethical concept, it seems to be necessary to give reasons why any set of such rules (which is almost certain to contradict the set of rules of some other country) may be *better* than any others?

Responding to the situation, Apel tries to find out whether fundamental elements of *communication* exist which might be used as a generally acceptable basis for global ethics. He claims that it should be possible to find such a basis which can be accepted by all partners of communication if this communication should work at all.

He subsequently analyses the *Act of Speaking*, and he discovers that any discourse mainly consists of situations in which a speaker uses language to *convince* others of what he thinks is *true*. Conversation thus includes the agreement on *logical rules of language* as well as the aspect of inter-subjective *consensus*. Apel further finds that succeeding or failing with such a discourse does not only depend upon whether the *present* partners in conversation agree or disagree, but that in addition any consensus must consider the aspect of *universality*. The discourse should not only aim to convince one special person, but it should as well be acceptable in terms of what Apel calls the *ideal community in communication* - beyond the present communication partners. In his view only this stance allows us to judge the *ultimate truth* of validity claims worked out in such discourse.

Apel and Habermas use these considerations as the starting point for their *Discourse Ethics* which is, first of all, to be understood as a *normative meta-concept*. *Discourse Ethics* has itself been developed by means of such discourse. The discussion on Discourse Ethics in detail has been going on for more than thirty years, within Germany as well as internationally. It tries to offer assistance in *co-operatively* working out required moral concepts for living and working together. The aim is, thus, to create communication in which it is possible to find consensus about genuinely relevant aspects of universal values, through global teamwork.

The concept, however, can be considered valid for intercultural co-operation only when, in a second step, every single partner in this conversation agrees to the results of the discourse. Of course, in international affairs, different partners have different attitudes and aims. Apel is aware of the dangers arising from conflicts of interests and power. But he argues that without the final aim of consensus, human action is purely destructive and thus, it is not acceptable to *everybody*. He demands that every decision in *Applied Discourse Ethics* should, thus, take into consideration not only the consequences for the partners involved in the discussion itself, but also the consequences for all those who are possibly in any way concerned. This is an important aspect of ethics in connection with science and economy. We may only mention here, e.g., the issues of global peace, and environmental and economic stability.

Following Apel, the *ideal community of discourse* is unfortunately hardly ever to be reached in everyday life. This circumstance now requires special consideration. Let us consider people in charge of others, e.g. politicians as representatives of their country, or managers who have to care for the survival of their companies. These people must then be aware of possible mistrust and strategic considerations of their opponents. It is not at all the intention of a responsibly-minded concept of ethics to ignore those dangers. The concept includes, therefore, not to follow such principles when others are then in danger of being left out of participation or even consideration. But it is also not to be accepted if only strategic and unfair actions of communication partners are the consequences of such caring attitude of one of the partners. Apels concept for *Applied Discourse Ethics* is thus: to use as little strategic planning and scheming as necessary, and as much orientation on the principles of the ideal discourse as possible.

As the first example, the application of these concepts to engineering action is described in the following section of this paper as it was discussed during the World Engineers Convention, Hannover, Germany, 2000.

5. INFORMATION AND COMMUNICATION: THE MEMORANDUM

The following *Trends, Challenges and Tasks of Information and Communication Technologies* are the results of the Congress on *Information and Communication* during the World Engineers' Convention 2000 in Hannover. They were discussed and agreed on, by the (about 800) participants of the Congress workshops. Here they are quoted in a condensed version:

"Global versus Regional Development: Law and Governance

The main trend today is towards globalisation: all communications and transactions take place within world-wide dimensions. To make the world a really global world, however, we need to reconcile national laws at the international level as well as mechanisms to enforce them, without succumbing to any dominant perspective. Hence we need to contribute to more political control of technology-triggered developments through making more information available to all citizens."

"Entrepreneurship on Different Scales: Economics and Business
Networking on the *global* "macro" scale is leading to both strong economic co-operation and mutual dependencies of large enterprises and countries. The liability and responsibility of global enterprises is no longer towards any specific country or people. There seems to be no control or governance possible through any single country. In parallel, *regions* are challenging traditional national politics

by developing their own political momentum. We need to create awareness for, and adapt policy to the joint design and implementation of technological, political and organisational renewal."

"Data Availability versus Data Security: Transportation and Processing of Data
Today, all information on the technological networks is available to everybody. But the misuse of the web and the breaking of data security are well known. Insufficient data reliability, trustworthiness and dependability are increasingly becoming a global problem. Data availability and data security are contradicting challenges on both the technical and organisational level. A new security culture needs to be developed concerning all developers and users. "

"Reality versus Virtuality: Acting within the Global Net
Within global networks we observe completely new ways of remote process control, and business transactions at a distance. Furthermore everybody is invited to experience personally the enrichment of life by such technology-based information and co-operation. This development is a big step beyond the *consumer* attitude of the traditional TV. But many experiences today are transmitted only through the technological networks. Thus they are frequently not accompanied by experiences of reality. Hence we need to design automation and control technology networks (e.g. power stations, factories, aircraft, etc.) so as to ensure that a competent human operator or system manager remains in charge of the system."

"Education: Technology-based and Traditional Education
The growing availability of information and knowledge allows new educational use of the web. All available knowledge must be easily and economically accessible for everybody on the web. Furthermore all education and learning should not only be based on technology and virtuality but also on traditional forms of education, on the reality of personal, social and professional life. "

"The Ethics of Multimedia Information and Internet-based Action
The calling-up and exchanging of information and pictures have proved their importance and necessity in personal life as well as in many fields of research, business, politics etc.. There is, however, the freedom of storing and sending all those pictures which symbolise the harmful or abusive side of human life (e.g. pornography, racism, violence and violent games etc.) It is unethical to transmit consciously and purposefully abusive, wrong and misleading information. Hence the individual responsibility of the engineer needs to be reinforced by a professional code (like the

Hippocratic Oath). Furthermore we need to discuss the ethics of information and pictures in view of the *cultural pluralism* of countries, their different traditions and value systems while avoiding to establish any *one value system* across the world." (Memorandum, 2000)

So far, the different aspects of engineering actions under global technology impact have been briefly presented here, as they were discussed by the World Engineers Convention, 2000. They demonstrate how engineers today are already aware of the needs and challenges put forward by *Applied Discourse Ethics* and the claims for universal validity of values implemented in engineering. Another example along the same lines of reasoning, is symbolized by the *Fundamentals of Engineering Ethics* suggested by the *Association of Engineers VDI*, Germany, in 2002.

6.   THE *FUNDAMENTALS OF ETHICS* AS SUGGESTED BY THE VDI, GERMANY

In 2002, the Executive Board of the *Association of Engineers VDI (Germany)* passed the new document *Fundamentals of Engineering Ethics.* They are intended to offer to all engineers, as creators of technology, orientation and support as they face conflicting professional responsibilities. Some paragraphs from this document are following here. They represent the thinking and wording towards *Applied Discourse Ethics* :

*From the **Preamble***

"Engineers recognise natural sciences and engineering as important powers shaping society and human life today and tomorrow. Therefore engineers are aware of their specific responsibility. They orient their professional actions towards fundamentals and criteria of ethics and implement them into practice. The fundamentals suggested here offer such orientation and support for engineers as they are confronted with conflicting professional responsibilities."

*Concerning the **Responsibilities** of Engineers*

"Engineers are responsible for their professional actions and the resulting outcomes. According to professional standards, they fulfil their tasks as they correspond to their competencies and qualifications. Engineers perform these tasks and actions carrying both individual and shared responsibilities."

"Engineers are responsible for their actions to the engineering community, to political and societal institutions as well as to their employers, customers, and technology users."

*Offering **Orientation** to Engineers*

"Engineers are aware of the embeddedness of technical systems into their societal, economic and ecological context. Therefore they design technology corresponding to the criteria and values implied: the societal, economic and ecological feasibility of technical systems; their usability and safety; their contribution to health, personal development and welfare of the citizens; their impact on the lives of future generations" (as previously outlined in the VDI Document 3780).
"
The fundamental orientation in designing new technological solutions is to maintain today and for future generations, the options of acting in freedom and responsibility."

"Engineers thus avoid actions which may compel them to accept given constraints (e.g. the arbitrary pressures of crises or the forces of short-term profitability). On the contrary, engineers consider the values of individual freedom and their corresponding societal, economic, and ecological conditions the main prerequisites to the welfare of all citizens within modern society – excluding extrinsic or dogmatic control."

"In cases of conflicting values, engineers give priority:

- to the values of humanity over the dynamics of *engineering*,
- to issues of human rights over technology implementation and exploitation,
- to public welfare over private interests, and
- to safety and security over functionality and profitability of their technical solutions.

Engineers, however, are careful not to adopt such criteria or indicators in any dogmatic manner. They seek public dialogue in order to find acceptable balance and consensus concerning these conflicting values." (VDI, 2002)

7. CONCLUSIONS: FUNDAMENTALS OF COMMUNICATION

We may derive certain basic rules for communication from these fundamental principles of communication and co-operation. They are reflected, e.g., in the VDI documents. The following agreements are essential for success of any communication and co-operation:

Partners on equal terms in communication
- have the same level of information on the subject of communication,
- are equally integrated into the whole process of finding decisions,
- freely offer and explain their opinions,

- know and accept the others' opinions *on equal terms* with their own opinions,
- are prepared to re-think their own positions taking into account new information or new aspects of opinions.

Such basic rules for communication are not directly part of Apel's and Habermas' concept of Discourse Ethics and thus, they cannot be found within their own works. But they can be understood as derived from the principle of the *Ideal Community of Communication*.

Thus questions of ethics, such as freedom of choice, international and intercultural acceptance, self-determination, and responsibility for humanity, peace, and our natural environment, are part of our everyday lives. They influence our decisions as actors within regional as well as global processes. It is the role not only of philosophers, but also of entrepreneurs, engineers, researchers, university teachers, students - of *everybody* - to take care of and deal responsibly with global peace, our environment, our partners, and the future generations.

The statements quoted in this paper have taken up this challenge, and the many different actors addressed are challenged to answer the questions of Ethics today. They are to play important roles in supporting human-centred thinking - be it in engineering or in wider societal involvement.

REFERENCES

Apel, K.-O. (1997): *Diskurs und Verantwortung (Discourse and Responsibility)*. Frankfurt am Main 1997.

Bundestag (2002): Interim Report of the German Bundestag. Enquete-Commission on Globalisation and World Trade, Bonn 2002.

Memorandum (2000): Information and Communication - The Memorandum. Professional Congress Information and Communication, VDI World Engineers' Convention, June 19-21, 2000, Hannover, Germany (http://www.zlw-ima.rwth-aachen.de/forschung/publications/the_memorandum.html)

Rose, C., Olbertz, E. and Brandt, D (2002): Global industry networking and the issue of ethics. Proc. 15th IFAC World Congress Barcelona, Spain, July 21-26, 2002, p 421.

VDI (2002): Fundamentals of Engineering Ethics. VDI, Dusseldorf, 2002 (http://www.vdi.de/imperia/md/content/hg/1 7.pdf).

www.elsevier.com/locate/ifac

THE ASSAULT ON IRAQ BY THE UNITED STATES—2003: ANTITHESIS OF DISPUTE SETTLEMENT?

J. G. Richardson

Decision+Coinmunication (Consultants)
Cidex 400, 91410 Authon la Plaine, France
decicomm62@aol. corn

Abstract: The challenges related to maintaining international system-stability were not met peacefully in the politically tense winter of 2002-3 when a predominantly American and British coalition prepared a military attack against Iraq's leadership under Saddam Hussein. The reason advanced for an assault was that Iraq's dictator, Hussein, had accumulated and deployed weapons of mass destruction contrary to sanctions imposed by the UN Security Council on Iraq. As threats of war intensified, non-US and non-British governmental and public clamour—round the world—condemned the resort to arms. War came on 20 March 2003, however, lasting 26 days. How did this happen, and were there options? *Copyright © 2003 IFAC*

Keywords: system, inside-outside pressure, stability, conflict resolution, dispute settlement

> ... the fate of all complex adapting systems in the biosphere— from single cells to economies—is to evolve to a natural state between order and chaos, a grand compromise between structure and surprise. (Kauffman, 1995)

1. KEY NATIONAL PARAMETERS, IRAQ
(Congressional Digest, 2002)

		Employment (1989)	Agriculture 43% Industry 26% Services 31%
Population (2001)	23.3 million	Health	Life expect. 67 yrs
Ethnicity	Arabs, 75; Kurds 20		Infant mortality 60/1000 (2001)
	Turkmen, Assyrian, others 5	Education	6 yrs compulsory
Religions	Muslim Shi'a 60	GDP (2001)	US$57 billion
	Muslim Sunni 35	Commerce (2000)	Crude-oil exports $21.8 billion to Russia, France, Switzerland, China
	Christian >5		
	Jewish-Yezidi >1		
Languages	Arabic Kurdish Assyrian and Armenian		Imports almost $14 billion from Egypt, Russia, France, and Vietnam
Political party	(single) Ba'ath Arab	Resources	Petroleum, natural gas, phosphates,sulphur
Area	437 072 sq km		
Capital	Baghdad	Literacy	58%

2. BACKGROUND

The stance of the United States in western Asia during recent decades stems from both the departure of Great Britain in 1967 from the Suez Canal region and the Arab-Israeli war of the same year. The U.S. subsequently

> «relied on its two principal allies, Iran and Saudi Arabia, to maintain the appearance of stability in the Persian Gulf. But 1979 undid that strategy. That year brought the overthrow of the shah of Iran, followed by the seizing of hostages at the US Embassy in Tehran and the Iranian Revolution led by Ayatollah Ruhollah Khomeini.» (Priest, 2003).

Also in 1979, the USSR invaded Afghanistan, which—because Americans recruited 'freedom fighters' for that country from far and wide in the world of Islam—proved to be another source of long-range difficulty for policy formulators in Washington, in both the executive branch and the Congress.

In 1986 the Goldwater-Nichols Act passed by the US Congress authorized a global structure of commanders-in-chief responsible for United States all-service coordination, planning, command, local liaison, and logistics. A 4-star general or admiral was authorized to command each of (now) five commands.

In 1998 ordinary citizen Donald Rumsfeld, a corporate executive and former defence secretary, signed a letter to President Clinton in a plea «to make removing Saddam Hussein and his regime» an objective of US foreign policy (Economist, 2003). This appeal was not acted upon during the Clinton presidency.

3. ANALYSIS

The large form in Rgure 1 represents Iraq (a chiefly Arab and, therefore, Muslim country) controlled by the Ba'ath Arab political party. The large inner circle represents the ruling Muslim component of the population. Conflicts, we know, frequently end in clashes between their antagonists, here *sides a* and *b*.

Inside the two small, intersecting circles lived *a*, Iraq's Sunni leadership from 1958 to 2003: 30%, comprising all those living within Iraq supporting the Ba'ath leadership, together with sympathetic elements abroad—including suspected relationships with Osama ben Laden's al-Qaeda. Inside *b* were the minority Shi'a, numerically superior at 60%, a population living in Iraq's southern regions. The *outsiders* within the overall pattern include the non-Arab Kurds, settled in the north of Iraq, together with other minorities and the many Iraqi abroad, all opposed to the Ba'ath hegemony. The large circle

inside the form is thus inhabited and led by the *insiders*.

The *outsiders*, on the other hand, are all those—including side *b*—who suffer at the hands of *a*, as well as parties external to Iraq who awaited a change of leadership in their native land. The constituents of the latter political impetus, or *b*, include the United States of America and a handful of other democracies; Iran; and the leaderships of certain other Arab or Muslim states in western Asia and elsewhere.

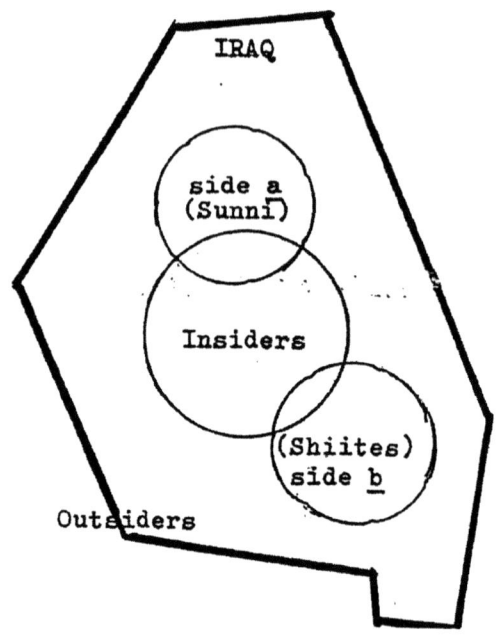

Fig. 1 Insider-outsider configuration, Iraq

The external component is known in conflict-resolution theory as the *third side* (Ury, 1999)—the first two sides being the *insiders* and the *outsiders*. The *third side* also included, during both the First Gulf War of 1990-1 and the Second Gulf War of 2003, members of the United Nations and especially its Security Council—the US and Britain included.

It is *the third side* that moved for definitive political, and then military, action against Iraq by 2002-3. The Amarican president, George W. Bush, made much publicy (including a speech before the UN in September 2002) of the accumulation of weapons of mass destruction (WMD) by Iraq's leader, Saddam Hussein, and Hussein's disregard of UN structures and inspections *in situ* to control such arms.

The U.S. also advancer the cause of bringing democracy to Iraq (and the region generally). The Americans, British and others generated, via the UN's Security Council, a second round of UN inspections (November 2002-February 2003); but the inspectors failed to find trace, except for a few indirect indicators, of the presence of WMD. «It's almost impossible to know if WMDs or suspect laboratories have ever existed..,» said biochemical-detection specialist Lieut. Valerie Phipps. «You can sanitize places so [that] you never know what was in there.» (Timea, 2003).

In the course of negotiations meant to resolve conflict and settle disputes, arbitration expert William Ury and others prescribe that the *third side* needs to make every effort to **avoid** the emergence of destructive measures, **resolve** the overt conflicts, and at the same time contain any struggle for escalating power by both insiders and outsiders.

To reach the stage of *avoid-resolve-contain*, the third side—the peacemakers—need, ideally, to be guided by the following.

Avoidance
a. Providing in ways that the population of Iraq can meet its need
b. Teaching, by helping people acquire the skills to deal with conflict
c. Bridge-building: forging relationships across conflict boundaries

Resolution
d. Mediating, reconciling conflicting interests
e. Arbitrating, determining disputed rights
f. Equalizing, or democratizing, power
g. Healing, by repairing injured relationships

Containment
h. Monitoring: paying attention to escalation
i. Being referee, or setting limits to fighting
j. Peacekeeping: providing protection

4. FROM THEORY TO REALITY

Tension in Iraq and its region failed to diminish from the time of the UN sanctions and inspections imposed in 1991-2. As Saddam Hussein first evicted the UN's inspectors in 1998 and then agreed to receive them a second time in 2002-3, his aura with side *b* and some outsiders deteriorated to that of «Saddam Hitler» or «Stalin Hussein», an *evil* force to deal with.

The situation worsened after (i) terrorist attacks on the United States in 2001 and (ii) American, with Allied support, retaliation against the Taliban in Afghanistan (2001-2002). **Avoidance-a**, above (*providing*), was taken in charge by the UN, but **Avoidance-b** (*teaching*) was not feasible, and **Avoidance-c** (bridge-building) seemed of no interest to Washington's leaders.

Resolution-d, -e and **-g** (mediating, arbitrating, healing)'were- of no priority to President Bush and his main advisors, while the American leadership envisaged **Resolutioil-f** (cfemocratizarion) as badly

needed «regime change» in Baghdad—increasingly and at any cost. The **Containment** phase (**h, i, j**, above) did not, indeed could not, apply in the face-off between Iraq on the one hand and the US-Great Britain coalition on the other hand, 2001-2.

After months of build-up, and speech after speech within the UN context and at numnerous national levels to dissuade American-British attacks on Iraq, air and ground assaults began simultaneously on 20 March 2003. Within 26 days Iraq failed as a state, and foreign occupants invested the entire country.

A word now on the metrics of this brief international war, whose mid-term and long-range sequels cannot be estimated at present writing with any assurance of exactitude.

The Americans and British committed a force of about 300,000 men and women against a roughly similar Iraqi number. Civil and military casualties as of 27 April 2003 included 30,000 Iraqi military killed, wounded or missing (according to Iraqi sources; Time reported more than 10,000 dead—*Timeb*, 2003); civilians killed, same source, 1,240 (2,000 from Project Body Count, a non-Iraqi NGO). On the Coalition side, the British lost 30 killed, the USA 126 killed, the Australian forces none. Non-combatant support troops from Poland, Spain and elsewhere suffered no casualties.

This unbalanced score, resulting probably most directly from a serious disequilibrium in opposing fire-power, typifies *asymmetrical* war (Leech, 2002a). In mismatched armed combat, the ingenuity of humans «further sharpen[s] the basic asymmetries of strategy and equipment (Leech, 2002b).

The financial cost of the war, borne chiefly by the U.S., was $20 billion, and each additional month of occupation comes to $1.5 billion. British costs have been, of course, additional, about $6 billion.

5. OPTIONS: THE «WHAT IF?» FACTOR

Analytical modelling and suppositions of a world otherwise, whose system stability is closer to equilibrium than was the deteriorating situation in and about Iraq between 1991 and 2003, require a few words concerning the «What if?» factor that frequently accompanies a review of history.

The main political players affecting involvement in the Second Gulf War were identical—albeit parallel—to those judging a peaceful outcome of the long Israeli-Palestinian dispute, especially in terms of demilitarizing the Second Intifada. The latter became the manifest polarization between the Islamic a non-Islamic worlds.

It should be noted, however, that during the period in question, the Palestinian leader, Yasser Arafat, for the first time named officially a second-in- command, Prime Minister Mahmoud Abbas. Could such an appointment, if made earlier, have affected attitudes throughout the Arab-Islamic world in terms of a less belligerent outcome of Israeli- Palestinia tensions ? *A reply*: this is possible.

The key actors observing the polarization mentioned above were the so-called quartet comprising the UN, the European Union (EU), Russia and the United States (with, more remotely in the geostrategic background, China). Among these four main elements the pressure for a war against Iraq came, however, from only one component, the US— supported most strongly by Britain, somewhat less by Spain, Italy, and a handful of Eastern European nations aspiring to membership in the EU.

What if the U.S. had listened, abroad and at home, to those opposed to armed conflict? *A reply*: In all probability, another roimd of UN inspections in Iraq would have followed, with possibly no clearer results, and the probability that the American assault would have been delayed—this time without British participation, given the weakened position of Prime Minister Biair's Labour Party in the face of strong public opposition in Britain to war in Iraq.

What if the U.S. had not used the pretext of a search for WMD in Saddam Hussein's Iraq? *A reply*: If such reasoning had been abandoned, the U.S. call for a *change of regime* in Ba'athist Iraq would have been weakened to the point of *no justification at all* for a pre-emptive war against Iraq.

What if the insistence of the U.S. unilaterally to bring democracy to the Iraqi people had been abandoned altogether? *A reply*: In this case, there would have been no *casus belli* whatever and— most probably— no armed warfare at all in Iraq.

What if Americn and British leaders had admitted to insufficiet knowledge of the WMD picture in Iraq? *A reply*: Their politial opponents would have debated and (very likely) thwarted warring action.

What if, finally, <u>any</u> country used pre-emption as a cause to attack another in order to effectuate regime change so as to install the machinery of freedom and democracy? *A reply*. Given the unquestioned justice of rule by law, the choice of unilaterally selected pre-emptive attack would result in non-linear relations between states: one of total disequilibrium, if not chaos—a situation barely conducive to regional or world peace.

6. A FEW CONCLUSIONS

• The United States decided, subsequent to the terrorist attacks against New York City and Washington in 2001, to change the regime in Iraq. The US considered this regime related to al-Qaeda and other extreme Islamic terrorism, while striving to search in Iraq for WMD.

• The use of diplomacy and its devices, mid- and long-term instruments of improved international stability, were forsaken by the Coalition *third side*.

• Given these two stances, the American leadership ceded momentarily, nevertheless, to world opinion by militating against Iraq through the UN and other intergovernmental fora, September 2002-March 2003.

• Given the same stance and the appeal to world opinion through the UN's systemic mechanism, American policy was maintained in order to pursue the elimination of Saddam Hussein himself as a «weapon of mass destruction»

• If the USA or any other nation pursues policies of this kind in the future, regime change may well occur through the application of asymmetrical force—but with little guarantee that rule of law, democracy and freedom would necessarily replace the regime overturned.

REFERENCES

Congressional Digest, (2002). Adapted from Vol. 81, No. 10 (<<^Disarming Iraq»), December, p. 291.

Economist, (2003)- «The Shadow Men», 26 April, p. 38.

Kauffman, S. (1995). At Home in the Universe, The Search for Laws of Self-Organization and Complexity, New York, Oxford University Press, 1995.

Leech, J. (2002a). Asymmetries of Conflict, War without Death, London, Frank Cass, 2002, 220 pp.

Leech, J. (2002b). Op. cit., p. xv.

Priest, D. (2003). The Mission, Waging War and Keeping Peace with America's Military, New York, W.W. Norton, p. 82.

Time (2003a). As cited in the Europe ed. of 12 May p. 37.

Time (2003b). «How Many Iraqis Have Died?» Time (Europe ed.), 21 April, p. 39.

Ury, W. L. (1999). Getting to Peace, Transforming Conflict at Home, at Work, and in the World, New York, Viking, 1999, pp. 3-24.

R. Fisher and S. Brown, (1989). Getting Together, Building a Relationship that Gets to Yes, London, Business Books, esp. «How Good is Our Relation- ship ?», pp. 178-9.

J.G. Richardson, (2000). «The NATO Campaign of Intervention in Kosovo on Humanitarian Grounds, and Its Aftermath», Proc. 7th IFAC Workshop, Ohrid. Republic of Macedonia, pp. 135-8.

ELSEVIER

IFAC
PUBLICATIONS
www.elsevier.com/locate/ifac

INFORMATION CONTROL.
ACTUAL PROBLEMS, EXAMPLES OF THE IMPLEMENTATION AND THE PERSPECTIVE

V.V. Kulba, V.D. Malugin, A.N. Shubin

V.V. Kulba - Professor, Head of Labortory of the Institute of Control Sciences;
V.D. Malugin - Professor, Head of Department of the Institute of Control Sciences;

A.N. Shubin - Professor, Vice-Director of the Institute of Control Sciences ;

Abstract: In the broad sense, information control is a mechanism where the control actions are of implicit, indirect nature and the information that is given to the controlled object has the form of informative picture, which is used by the object to work out an allegedly autonomous line of behaviour. Information control exists for a long time, but until now, it remained ancillary and insignificant as compared with other methods of control. In the second half of last century, the role of information control began its dramatic growth. The paper is devoted to discuss the causes of this phenomenon. *Copyright © 2003 IFAC*

Keywords: information control, audience, cognitive, media planning

The present report deals with control of society, groups, and social structures. By control of society is meant the method of influence which incites the people to act orderly, execute the required actions, and observe the laws. Four main methods to manage society, groups, and individuals are known: administrative (organizational), economical, socio-psychological, and legislative methods.

Analysis of using the traditional methods of control reveals that they are rarely used to act directly on the broad masses. At the same time, the increasing role of the popular masses in social and political life drew attention and activity of the ruling forces to the J r methods of direct centralized influence on the society.

Being aware of the historical irreversibility of appearance of common people in .the social life and "high" policy, the ruling structures try to gain their ideological and political support. They pose the problem of forming public opinion expressing the standardized opinions and estimates in compliance with the interests of the ruling stratum. Such a method of centralized control, the so-called informative J control, exists for a long time. Previously, it was unnoticeable and played an ancillary role. Only now, with the advent of the information technologies, its role started to grow in importance.

By the information control (IC) is meant the process of working out and realizing managerial decisions in an environment where the managerial actions are of implicit, indirect nature and the information that is given to the controlled object has the form of information picture which is used by the object to work out an allegedly autonomous line of behavior. Therefore, the aim of information active is to ensure the desired behavior of the controlled object as defined by the subject. For example, in the advertising business they make use of the AIDA formula, which is the acronym of Attention, Interest. Desire, and Action, that is, the task of information management is not only to inform, but also to recall, convince, and consequently, manage. It is precisely by means of this method of acting upon human minds that the ruling forces expect to subordinate public opinion to their influence on an unheard-of scale. The authorities regard the mass media an efficient tool enabling them to clothe the monitoring and management of the popular masses in new forms.

In this sense, of interest is an observation of President Richard M. Nixon that it is much more profitable to invest a dollar in the mass media, in propaganda, that ten dollars in designing new weapons. He substantiated this statement by that the probability of using weapons in the modern world is small, whereas propaganda operates every day, every hour. Albert Speer said "Eighty odd million people were deprived of independent thinking and subjected to the will of the man by means of the technical facilities such as radio and loudspeakers".

The mechanism of information action is based on manipulating human minds by purposefully introducing information, reliable or corrupt. The secret of information control lies in professional use of the entire spectrum of human weak points, "human likes, in applying masterly the methods for stimulation of numerous illusions" based on human activity and imagination. Competent information action results in instinctive, automated choice whose causes are enrooted by this action in the subconscious and conceal themselves in the depths of human psyche. Manipulation of conscious is based on switching off the common sense and logical thinking and then acting upon the subconscious senses, and imagination.

Methodological basis of the informative control relies largely on the ideas of Italian political theorist Antonio Gramsci. He asserted that in order to each strategic aims one does not need to do head-on changes in the social basis, but rather use the intellectuals to perform a "molecular aggression" through the superstructure against the social conscience and destroy its cultural nucleus. By the way, this statement is consistent with the fact that evolution of the human conscience can be attained much easier that its revolutionary changes. This kind of management aims at forming the message about the real situation so that, despite its inadequacy, people take it as a matter of course and act appropriately.

The results achieved in attaining the IC aims can be classified as follows:

• changes in the degree of information and volume of knowledge on certain issues and impulse to create new information through the human cogitative potentiality (the so-called cognitive effect),

• changes in and/or shaping of convictions, opinions, interests, standpoints, and points of view,

• formation of the action effect related with a wide spectrum of practical actions such as voting, participation in rallies, strikes, demonstrations or, on the contrary, nonparticipation and so on,

• formation and/or modification of psychological norms, characteristics, and so on.

Therefore, the main purpose of IC is to act upon human conscience with the aim of endowing it with the desired properties and qualities that approach to the ideal as seen by the authorities. It is only natural that the mass media are not the only system operating in this field. Society has a branched system of educational and cultural institutions closely related with the mass media. There exist diverse kinds of problems that can be tackled using IC. They include strategic problems I such as change of social system, stimulation of the general public to respond more actively to some strategic events, elections, attitude toward other states, points of view, nations, parties, and external and internal events, competition with rivals, as well as ordinary advertising.

The following main forms (functions) of the informative control by the mass media are differentiated:

1. informing function which ensures the human rights to information, support of the informative potential required for stable social development, and so on;

2. function of organization of behavior which forms certain rules (algorithms) of behavior of the members of society whose its special cases are exemplified by upbringing, education, and creation of the "desired emotional and psychological disposition, and

3. function of communication which establishes and supports relations between individual segments of the audience, as well, as establishes the feedback for estimating 1C efficiency, informative action being constantly updated through the feedback until the aim is attained.

The mass media play the role of competent, although very specific, managerial organs. Formally, they do not make managerial decisions. Their main objective is to exercise indirect ideological, psychological, and other actions upon the social processes; actions that do not have the form of law, order, and so on. The aims of these actions are defined by the leaders of the state or regions, rather than by the management of the mass media. Yet, the mass media virtually enjoy tremendous independence in managing the society. This is due to the tact that, they themselves generate scenarios of 1C realization, as well as texts and accompanying materials using various technical facilities such as "amplifier" - provides the society with information required to the subject; here, a smart form of its presentation perturbs social balance, changes the lifestyle, and increases in the individuals the sense of unfitness of their behavioral schemes until this phenomenon becomes collective and involves the major part of society, and "prism" - similar to light refraction in the prism, each mass media records, details, and transmits the desired tendencies clothed in a simple understandable form with the attributes of the everyday life of each individual, and so on.

The following technological stages can be identified upon realizing the informative control:

• realization and formulation of the problem,

• analysis of the situation,

• formulation of the aim, construction of the goal tree, development of the program of realization of 1C;

• segmenting of the audience, study and analysis of the characteristics of segments of the target audience,

• development of concept, strategy, and scenario of realization of the informative control,

• development of plans of actions, choice of methods and means for realization of informative control,

• realization,

• monitoring and checking, correction, and

• efficiency analysis.

Stated differently, the mass media must constantly exercise informative action upon the society by executing strategic, tactical, and operative psychological actions. This makes the difference between the informative management and informative warfare (confrontation of two or more systems). This difference can be clearly seen in the paper of Winn Schwartau at the congress on informative warfare held in Brussels in 1996.

High selectivity of action and virtual absence of boundaries (international level at the limit) concreteness and expeditiousness, fast restructuring of the methods and means of actions depending on the varying environment, the possibility to concentrate quickly efforts on one or another object, region, social group, the possibility of using jointly different methods and means of informative management, and

comparatively low cost of developing and realizing the j managerial decisions with high efficiency of their introduction into conscience can be cited as the advantages of informative management.

REFERENCES

Kulba, V.V. (1996). On Informative Control. In: Inform. Vychisl. Techn. [in Russian]. **Vol.1**, pp.259-263. Nauka, Moscow.

Kulba, V.V., Malugin V.D., and Shubin A.N. (1997). Informative Control (Background Methods, and Means) [in Russian]. IPU RAN, Moscow, p.54.

Kulba, V.V., Malugin V.D., and Shubin A.N. (1997). Informative Control (Background Methods, and Means) [in Russian]. IPU RAN, Moscow, p.54.

Zverintsev, A.B. (1997). Communication Management [in Russian]. Soyus, St. Petersburg, p.287.

B.A.Grushin, and L.A. Qnikov (eds.). Mass Information in an Industrial Soviet Town. An Essay of Sociological Study [in Russian]. Izd. Pojit. Liter. 1980, p.444.

Kara-Mursa S. (1999) We ourselves dug our grave. [in Russian]. Nash Sovremennik, **Vol.6**, p. 188-204.

ELSEVIER

IFAC
PUBLICATIONS
www.elsevier.com/locate/ifac

WAITING FOR THE INTERNET DHABA: CASE STUDIES OF TECHNOLOGY ACCESS, INDIGENISATION AND E-GOVERNMENT IN INDIA.

David Smith

*School of Art, Media and Design, University of Wales College, Newport,
PO Box 179 Newport, Wales, NP18 3YG UK*

Abstract: This paper examines a number of case studies in development towards E-government and citizen participation emerging from the EU-India Cross-Cultural Innovation Network Project, in which the author has been a partner for the past four years. The paper makes a case for re-examination of dominant technology-centred 'globalised' models of development and points to the need for "indigenisation" of ICTs to provide tools for citizen participation. It is suggested that there is no single, simple answer to this problem, but that one important component of the solution lies in a holistic and Human-Centred approach to Information Design. Attention is drawn to Indian arguments regarding the importance of a bottom-up approach (the so-called "Subaltern Trajectory") in the evolution of both E-Government and E-Governance. The paper argues that the 'Subaltern Trajectory' has already produced a number of sustainable exemplars for citizen-centred service delivery in the context of E-Government, but that the work of moving towards E-Governance is only just beginning. *Copyright © 2003 IFAC*

Keywords: Human centred design; Knowledge Transfer; Knowledge Engineering; Rural Development; Developing countries

1. INTRODUCTION

The advent of the so-called "Information Society" has been heralded for almost as long as we have had practical personal digital computers. Predictions of the nature of this society have ranged from gushing euphoric determinism to the juvenile dystopias of the cyber-punk genre. Most of the more balanced views, however, agree that the new techno-social order is likely to be characterised by three elements:

- The intensification of a systematic exchange and mediation of digitised information;

- A rapidly increasing part of the GNP generated through the processing of information;

- New opportunities for the economic, social and cultural development of modern society.

This paper is concerned with the last of these elements, namely with "social and cultural development of modern society", and most particularly with the implications of convergent Information and Communications Technologies (ICTs) for civil society. This field of concern has come to be called "E-Government" or "E-Governance". There are, in fact, critically important distinctions between the implications of the two terms, and this will be examined below.

Although there has been considerable discussion concerning E-government in the EU and elsewhere in the so-called "developed" world, the majority of proposals for the development of E-government have tended to be framed in terms of "how do we get information about this or that government policy or action on the Internet?", or, more recently, "how do we provide this or that government service on-line". In fact, considerable progress has been made in some areas. Local and national governments throughout the world (and not only in the 'North') have enacted laws related to freedom of information, and this has been reflected in the widespread development of web sites and other on-line sources of previously inaccessible information. The British Government, for example, has set ambitious targets (which are unlikely to be met!) for the digital delivery of all government services by 2005.

This is, of course, no bad thing in itself. We should welcome any application of ICTs which makes the workings of government more transparent, or which allows us to enjoy with the minimum of difficulty the full range of services to which we are entitled as citizens. Though of course we should be careful not to deceive ourselves. Cairncross (1997) pointed out that "...the use of [ICT]... will not help citizens root out facts that a government is determined to keep from them." Indeed, as she went on to argue, "...although more people may be able to read it, [information]... it will not necessarily be more useful or more comprehensible simply because it is electronically available..."

These are important caveats. Successful E-Government depends on the probity of those who operate it and on the educational levels (and sheer determination) of those who attempt to use it. Recent surveys suggest that many US citizens distrust their government on-line just as much as off-line. Ironically, whereas 70% of those sampled were willing to give personal information to a commercial web site to obtain a product or service, only 29% would do so to a government site. This is matched by an equal lack of enthusiasm in the UK, where usage of government sites and on-line services has not grown as predicted. It seems that a significant percentage of citizens in both countries simply do not trust their governments.

E-Government also, of course, depends on accessibility. Much has been written in recent years regarding the so-called "Digital Divide". The concern that a significant part of any population may not be able to access on-line services and/or information as a consequence of poverty or lack of basic infrastructure has become a constant recurring theme, and

governments have dedicated substantial resources to providing as many citizens as possible with access to the Internet. The provision of public access to high-bandwidth telecommunications has become a major plank in government responses to the perceived problems of the digital divide. Whether or not this is missing the point will be discussed below, but it is worth noting that even in the USA, only 68% claim to have access to the Internet at home, at school or in the workplace (Washington Post, 2003).

These problems are tractable, even if the solutions will be rather difficult to put in place. But even when governments are finally successful in solving them, the implicit political rhetoric of the dominant approach to E-Government will still impose an obstacle to the "social and cultural development of modern society" referred to above. The bulk of development in this field has been focussed on what might be termed the "Citizen as customer" model. The tenor of much of the discussion appears to set the entire relationship between government and governed in terms of the provision and consumption of services, which can, of course, equally be provided by commercial agencies. When the State of California defines E-Government as "...electronic, efficient and easier to use government...", this can be seen as little more than the extension of E-Commerce into the realm of civic society, and its ultimate vision is one of quasi-commercial efficiency, rather than one of evolving the democratic process in the light of new possibilities for action.

This vision of E-Government provides leverage for those commercial corporations which have already invested heavily in the technology of E-Commerce. The redefinition of government as a special instance of commerce allows the "repurposing" of available expertise, so long as the debate can be conducted in appropriately restricted terms. Thus, the management consultancy 'Accenture' (2001) argue that:

"...only in isolated cases has online service delivery moved all the way up the maturity curve, enabling citizens to complete entire transactions with government, such as paying taxes or claiming benefits online..."

Nobody would argue against the proposition that these represent useful, if limited, advances in service delivery, but they cannot by any stretch of the imagination be represented as moving "all the way up the maturity curve"!

This kind of thinking also pervades many proposals for technological innovation supposedly carrying us into the information society. The influential Bangemann Report "Europe and the global information society"

(1994), for example, whilst proposing priority demonstrator ICT applications which would blaze the trail and "launch the information society", could still only offer an essentially technocentric action agenda:

1. Application One: Teleworking
2. Application Two: Distance learning
3. Application Three: A network for universities and research centres
4. Application Four: Telematic services for SMEs
5. Application Five: Road traffic management
6. Application Six: Air traffic control
7. Application Seven: Healthcare networks
8. Application Eight: Electronic tendering
9. Application Nine: Trans-European public administration network
10. Application Ten: City information highways

These are all very worthy projects, but they are redolent of a "data society", not an "information society". There is nothing here to challenge or reposition the relationship between citizens and state. Nor is it intended to do so.

A few conscious dystopias aside, there is inadequate mature discussion of the potential trajectories for the evolution of civil society beyond a general deterministic assumption that an "Information Society" will necessarily be benevolent. We simply cannot assume that this will be so. When Al Gore, for example, spoke of "...forging a new Athenian Age of democracy...", it was left to Cairncross (ibid p. 259) to point out that "...Athenian democracy excluded women and slaves, a majority of the population, from the rights of citizenship, and the Athenian assembly was notoriously prone to being hijacked by oligarchs and demagogues..."

ICTs may indeed allow for the direct control and management of human services such as health, welfare, legal services and social security, but this process may also weaken and even replace the intermediary agencies such as voluntary groups, associations and community centres which have performed the role of filters and analysts of information and mediators on behalf of citizens for dealing with the bureaucracy. While electronic media certainly provide the channels for new forms of democratic participation, there is also a risk of further social divisions along fault-lines which we do not as yet even suspect. We need to innovate and experiment as never before. And we must reject once-and-for-all the "one best way" implications of the E-Commerce model. We need instead to evaluate the lessons that are emerging in other societies. We need also to consider not E-Government, but rather E-Governance: not "Citizen as Consumer with periodic voting rights" (in which electiuons become ever closer to commercial brand competition in a commercial market), but rather "Citizen as active participant in a new form of democratic civic society".

2. LESSONS FROM INDIA: THE EU-INDIA CROSS-CULTURAL INNOVATION NETWORK PROJECT

The EU-India Cross-Cultural Innovation Network project, running from 1999-2003 provided a frame of reference in which it was possible to examine the networking of knowledge within and between different cultural and subcultural groups. The project was concerned with the fostering of proactive collaborations in applied research in socio-economic and entrepreneurial innovations through international academic and entrepreneurial networking (Brandt, 2003). It brought together academics and social and economic entrepreneurs from Europe and India to undertake critical evaluation and collaborative dissemination of innovations. One focus for attention was knowledge networking and new models of civic society.

The development of E-Government and E-Governance in India takes us out of the frame of the globalised Western business model, and allows us to explore some of the fundamental issues in developing culturally appropriate participant democracy using new technologies. As the world's largest democratic state, India offers a richness of scale and diversity unparalleled anywhere in the world. Paradoxically, India is a village-based agrarian society which is at the same time a world centre of excellence in IT and one of the world's greatest industrial economies. With 1000,000,000 people speaking 850 languages and dialects, India offers an arena for the development of models of E-Government which are based in and on diversity and not necessarily predicated on a triumphalist view of Western market capitalism.

3. THE 'SUBALTERN TRAJECTORY' AND THE 'DIGITAL PROVIDE'

About 70% of the population of India is spread over around 700,000 villages, some of them situated in remote areas with little or no direct road access. This has meant that to conceive of the problem simply in terms of providing an expanded communications resource is to completely miss the point. According to Jain (2003), attempts to bridge the "digital divide" by providing the infrastructure to enable a one-way flow of information from government to citizens (so-called G2C solutions) will inevitably fail to address the real needs of ordinary people.

Jain distinguishes between two distinct ICT trajectories which are currently active in India. The first is the conventional ICT engineering model associated with central government initiatives and involving conventional science and technology institutions and organisations. This is an essentially 'top-down' technocentric process, aimed at providing solutions to perceived administrative needs or to 'digital divide' issues. Political and/or bureaucratic patronage means that such initiatives often achieve a high profile. Some of these developments are, of course, very successful: for example, computerisation of the land registration procedures in the state of Andhra Pradesh, has helped to minimise corruption in a notoriously corrupt area of public administration. But it is the contention of this paper that this is still working within the 'Citizen as Client' model.

On the other hand, however, there is a parallel but less visible trajectory which applies the technological capabilities of smaller groups and Non Governmental Organisations (NGOs) to widely distributed and unorganised populations in rural and semi-urban areas. As Jain points out, there are many examples of such initiatives, though they are often ignored in mainstream discourses on Indian ICT developments. Furthermore, the rhetorical language of the 'digital divide' neglects the rich knowledge resources of the so-called "information poor". Jain argues that laying down cables and providing access to computers and the Internet to enable a one-way flow of information is an inadequate strategy. He proposes instead that the aim should be to provide digital support for the networking of knowledge. For him, the ideal strategy is to establish two-way communication between those who have access to ICT services and those who do not as yet have it. He downplays the value of attempting to bridge the 'digital divide' and speaks instead of the "Digital Provide."

The principle of the Digital Provide can be explored through a examples studied in the context of the EU-India Cross-Cultural Innovation Network Project. Chakravarthy and Jain (2003) describe sixteen case studies and five 'vignettes', ranging from major state-level rural intranet projects (the Madhya Pradesh 'Gyandoot' project) to small-scale local knowledge clusters. Gyandoot has become the largest rural intranet project in India, and perhaps in the world.

The common feature in most of these projects is the locus of delivery. Many public access projects in Europe and the USA depend on private ownership of computers or access to them via libraries, schools or the cyber-café. Despite rapid growth in private computer ownership, however, India does not have such a resource infrastructure, and although the number of cyber-cafés in India has doubled in the past two years, there are still only 50,000 of them throughout the country. Most of these are located in wealthy middle class areas of major cities, and very few rural Indians

ever come near them. Even fewer could afford them in any case. What is needed is a new resource, not a culturally inappropriate globalised implant, but a truly Indian institution – the cyber-dhaba, perhaps, based on the village dhaba (teashop and café). Encouragingly, that is exactly what is beginning to emerge.

The key problem for any E-Government anywhere is sustainability. Central funding is all very well, but it inevitably comes to an end. The long-term viability of projects such as Gyandoot depends on what happens after the umbilical cord is cut. There is no prospect of periodic central investment in these projects, which must instead develop into profitable businesses to survive. This is where the "digital provide" model really comes into its own. A case study described by Chakravarthy and Jain (ibid) illustrates this very well.

Drishtee (Hindi: 'vision' or 'light') is an innovatory Government to Citizen (G2C) model which seeks to make a paradigm shift by serving villagers directly through ICTs, rather than through civil servants. Drishtee installs low cost community-based rural intranet projects in villages or groups of villages. A local villager runs a kiosk, or 'Soochanalaya' (PC, modem and peripherals), which is financed through a government-sponsored loan. The idea is to run the kiosk as a revenue-generating business. The owners pay back the loans from their earnings.

Typically a kiosk serves 25-30 villages (20,000 – 30,000 people). The person operating the kiosk is not a state employee but a private entrepreneur. He or she runs the soochanalaya on commercial lines on the basis of a contract with Drishtee. The operators do not receive a salary but instead pay 10% of their income as commission to Drishtee. They also bear the costs of stationery, maintenance and all other bills. A typical business plan for a kiosk expects the owner to be earning at least 6000 Rupees (Rs 50 = approximately $1 US in 2003) by the third quarter of the first year of trading. There is growing evidence that this is both attainable and sustainable.

Mansingh et al (2003) examined the economics of village kiosks in the Punjab. They argue that for financial viability, these centres must become multifunctional. Apart from Internet access and purely ITR-related services, they may also engage in training provision or the repair and maintenance of electrical and electronic equipment. Some carry out data processing services to schools and local councils ('panchayats'). The core business becomes the nucleus for a whole series of related enterprises, and each kiosk typically gives direct employment to 3-4 persons and may provide indirect employment for several others. Mansingh et al (ibid., p. 98) cite a report from the All India Society for Electronics and Computing

Technology (AISECT) based on analysis of 2,500 centres in 24 Indian states which suggests that this model is economically viable even today. This is all very encouraging, but the final breakthrough will only come when telecommunications regulations are amended to permit the ubiquitous PCOs (Public Call Offices) to become effective digital communications hubs.

4. CONCLUSIONS

It has become conventional to assume that the public sector has only a limited role to play in the 'Information Society'. Leadbeater (2000), for instance, has compared the apparently remorseless pace of change in the private sector with a lack of innovation in the public domain and concludes that our ability to innovate in our civic and public institutions is diluted compared with our ability to innovate in our commercial, technical and scientific areas of life. But we must beware of drawing the false conclusion that this is either universal or inevitable. The public sector in India as elsewhere has shown little enthusiasm for providing infrastructure for the poor. In the end, the market alone will make global citizens only of those who can afford it.

The solution for India, and the lesson for the rest of the world lies in the fine balance between the local and the global which is achieved by the best of the E-Government initiatives current in India. With a few exceptions, almost all of the initiatives studied in the course of this Project were launched by national or state government organisations for reasons of efficiency. With minor exceptions but again, almost without exception, they have been taken up and given value by small-scale public initiatives combined with the entrepreneurial flair of private individuals. One immediate benefit has been in the reduction of the opportunities for corruption and irregular charges which were formerly part of routine administrative practices. Another has been to eliminate the need to travel long distances to obtain simple routine documents such as birth certificates etc. But much more importantly, the concept and growing practice of knowledge networking has placed the ordinary citizen in a quite different relationship to the power structures of administrative hierarchies.

This is where the move to a new citizenship really begins: "[E-governance] is not just about service delivery over the Internet. It is not just about how digital access to government information or electronic payments. It will change how citizens relate to governments as much as it changes how citizens relate to each other. It will bring forth new concepts of citizenship, both in terms of needs and responsibilities. E-governance will allow citizens to communicate with government, participate in the government's policy making and communicate with each other." (Web site of the Inter-American Development Bank)

Nobody pretends any of this will be easy, and the developments in E-Government discussed above are, of course, still a long way from providing viable frameworks for, much less proven models of E-Governance. But they provide us with some important lessons which we must now take on board.

5. REFERENCES

Accenture (2001) *Rhetoric versus reality - closing the gap.* Corporate Brochure

Bangemann M. (1994): *Report prepared for Europe and the global information society: Recommendations to the European Council.* Council of the European Communities, Brussels www.ispo.cec.be/infosoc/backg/bangeman.html

Brandt D., Ed. (2003) *Navigating Innovations: Indo-European Cross-Cultural Experiences. Volume 2 Shaping Information & Communications Technologies for Regional Development.* India Research Press New Delhi.

Cairncross F. (1997) *The Death of Distance.* Orion, London.

Chakravarty S. & Jain A. (2003) Informatics, Communication and Development: 21 case studies from India. In *Navigating Innovations: Indo-European Cross-Cultural Experiences. 2 Shaping Information & Communications Technologies for Regional Development.* (Brandt D. ed.) pp. 161-238. India Research Press New Delhi.

Inter American Development Bank web site (2003) www.iadb.org/sds/itdev/governance.htm

Jain A (2003) From Digital Divide to Digital Provide: the subaltern ICT stream in India. In *Navigating Innovations: Indo-European Cross-Cultural Experiences. Volume 2 Shaping Information & Communications Technologies for Regional Development.* (Brandt D. ed.) pp. 149-160. India Research Press New Delhi

Leadbeater C (2000) *Governing the Weightless Society.* www.govtech.net/publications/visions/nov00vision/charles.phtml

Mansingh A, Sekhon HS, Dron J. & Rajapillai V. (2003) Innovations in IT Education in India. In *Navigating Innovations: Indo-European Cross-Cultural Experiences. Vol. 2 Shaping Information & Communications Technologies for Regional Development.* (Brandt D. ed.) pp. 77-99. India Research Press New Delhi.

Washington Post (14th April 2003) *Survey finds Americans split on 'E-Government',* p. A14 See also www.Washingtonpost.com

GLOBALIZATION AS A KIND OF LAW OF HISTORICAL DEVELOPMENT OF THE WORLD COMMUNITY: MERITS AND DEMERITS

A.N. Shubin, V.V. Kulba, V.V. Tsyganov

Institute of Control Sciences Russian Academy of Sciences
Profsoyuznaja Str., 65, Moscow, 117342, RUSSIA
E-mail: shoubine@ipu.rssi.ru.kulba@ipu.rssi.ru, bbc@ipu.rssi.ru

T.I. Ovchinnikova

Moscow State Institute of Steel&Alloys, Leninsky prospect, 4, Moscow, 117936, RUSSIA,
E-mail: tatianka_21@mtu-net.ru

Abstract: The globalization processes are under control of the international finance&information oligarchy. In the nearest future that may result in global totalitarian net. Ownership of the capital&information and control of its flows become today main tools of total control and global oligarchy domination. *Copyright © 2003 IFAC*

Keywords: global, oligarchy, control, information, capital, flows.

1. INTRODUCTION

Human civilization as a whole developed irregularly and passed the periods of the integration and disintegration. Technological changes provide the progress in the communications. New integration wave is provided by capital&information flows, telecommunications, global economy nets. It produced interdependence and global unity of the social systems.

From the other side, most of the countries in the world have no possibility for self-adaptation to these changes. Now the globalization processes are under control of world finance&information oligarchy. Its policy provides rupture between rich and poor countries. That may provide global totalitarian net. World oligarchy mastering capital&information, and control of its flows turn into main instruments of total control and world supremacy. Globalization processes advance on traditional world often in the form of aggression. Societies, political institutions, leaders and citizens most of countries are worry of perspectives of the possibility to lose the independence. Many countries must take a measure to defend their cultures, traditions and values from the alien information influence, and from the evident expansion mechanisms. Thus survival of the mankind needs joint efforts.

2. GLOBALISATIONS CYCLES

All history of development of the world community testifies that processes globalization developed together with the world community going up and down from time to time. The well-known examples of globalization processes are propagation of Christianity and Communism ideology. Forms of display of this process can be seen both on early, and at late stages of development of community, including the following forms: struggle for vital space – capture of the another's and not mastered territories, colonization and disintegration of colonial system of the world, local&world wars etc.

Last two centuries of a human history have passed in an atmosphere of progress technogene civilizations. Its major attribute is the accelerated scientific and technical progress inducing changes in all spheres of social life. The technogene civilization was preceded historically with the first type of development for which the slowed down rates of social changes were characteristic, domination of traditions and religious-mythological thinking.

The termination of 'cold war', finished in 1991 with disintegration of the USSR, has resulted in disappearance of bipolar system of the device of the world, that in turn has sharply sped up formation of the global market. The structure of the world economy again has got multipolar character.

The big unevenness in a standard of living and development of the countries is the main source of contradictions on the Earth. The several states have achieved such level of technological development and demand such huge consumption of resources which already exceed real opportunities of a nature. At the same time many peoples till now are practically discharged of elementary education, suffer from famine and deprivations.

The USA, left the winner in a global antagonism with the USSR, aspire to fix geopolitic results of the victory. The American economy became the main locomotive of development of the world economy. But, in one and a half or two decades China can overtake USA if the Chinese economy will keep present rates of development.

The global system of capitalism represents a circuit of trading platforms interconnecting on the base of modern communications in different regions of globe. It practically instantly allows to make remittance milliard means from one region of the world for another one.

All system of global capitalism is in a condition of dynamic balance and any indignation can result it in a dynamic nonequilibrium condition, and then in crisis.

The main distinctive feature of modern system of the international relations consists in formation and development of the powerful regional coalitions allowing more effectively to protect their interests on a world scene. Formation of the market goes through development of economic regional associations – the North American Free Trade Agreement (NAFTA), the European Union (EU), Asian-Pacific Economical Cooperation (APEC) etc. On their share it is

necessary approximately 80% of world gross national product, 82% of all state budgets of the countries of the world and 85% of export. Probably the global market in XXI century will be formed on this basis. The central place in the international economic system is played with 'the Big Seven', and also the World Trading Organization created under the initiative of USA.

'The Big Seven' under the control of the USA carries out the informal control above the World Bank and the International Monetary Fund which play a key role in formation of global financial and economic system.

The beginning of new century falls on a critical stage of a history of mankind. The population explosion the nearest 20 years will result in sharp growth of the population of the Earth (under forecasts, approximately up to 10 billion person in 2020 which big half will live in the Asian region. Intensive industrial development of several countries which have been in the round of countries of the third world and rising growth of consumption in the advanced countries sharply will aggravate a problem of resources. Already today on a share of 5% of the population of the Earth living in USA, it is necessary approximately 40% of world power expenses and over 60% of emissions in an environment.

The idea globalization testifies to an aggravation of problems of survival of mankind and understanding of necessity of joint efforts on their overcoming. Certainly, growth of interdependence in the field of economy and the finance inevitably will result in changes in sphere of a politics and culture. On a background of process globalization movement of peoples on search of ways of harmonious association of mankind is observed. The mankind should be united on the basis of the coordination of interests and interpenetrating values of the worlds coexisted today. And, despite of aggressive efforts of supporters of the unipolar world, globalization is destroyed by internal processes of self-preservation of different cultures. Today the vector of development of a society should turn in a direction of high quality maintenance of life of all of its members.

3 .ADAPTIVE CONTROL PRINCIPLES

Chestnut and Kopacek (1989) indicate that to reduce the risks and likelihood of the international instability it is possible to apply adaptive control principles for international conflict resolution. Adequate adaptive procedures provide identification of controlled object structure and parameters of environment and, finally, generating the controlling actions making use of current information, obtained from the elements, in order to achieve the optimal state of the system as a whole.

Technological changes provide the progress. But these changes produce also gap between rich and poor countries. Many poor countries of the third world have no possibility for learning and adaptation to these changes and need the support.

The technological changes are created by high technologies (high-tech). Advanced technology produces both positive and negative effects of social stability. First of all, the technically advanced systems that are in existence, as well as those being built, can be used effectively to help people to realize the benefits that are possible with the present-day high technology. From the other side, potential of high-tech now is accumulated in a few major leading countries. Multinational corporations from these countries supply high-tech all over the world. To provide new high-tech it is necessary to realize R&D, know-how etc. That needs appropriate intellectual and financial resources. The prices of high-tech goods and services are mainly not competitive prices because of the monopoly on the results of R&D and know-how. From the other side, price of traditional goods, provided by developing countries, are under strong competitions. For this reason investigations and investments in high-tech are in most of cases preferable.

Mechanism of global economy functioning is arrange by world finance&information oligarchy in such a way that financial and intellectual resources drain from developing to leading countries. The result of the capital and brain drain is the lack of financial and intellectual potentials in developing countries. It becomes more and more difficult to provide R&D, know-how and new high-tech. Therefore technological development produces economical and social rupture between leading and developing countries. That provides likelihood of social crisis and instability, and as a consequence tensions between different countries. Corresponding to the well-known business magazine "Euromoney", edited in UK, in 1999 almost one half of 180 countries in the world (85 countries in March and 87 in September) had credit rating, equal to zero in the same periods. More the one half of all countries (92 from 180) had no access to international bank finance.

Developed countries provide some efforts to diminish this gap and to improve international stability. The third world support should correspond to the level of this gap between rich and poor countries in a way used in adaptive control systems. From the other side, systemic investigation of the problems dealing with the design of the adaptive systems to control organizations, including as the main component analysis and the account of the human factor effect. It should be understood as an activity manifestation of the people or collectives (elements of control systems) is caused by the availability of their own aims, not necessary coinciding with the goal of the system in its entirely (Tsyganov, 1990). Such elements may utilize available information channels connected with the system control center in order to improve the current or future state. From the other side, "active" people deals with the process of this support may predict the results of the adaptive control procedures and to use that knowledge to reach their own aims. In fact, in many cases support given from developed to poor countries had no effect because of the failure of the mechanism used (corruption, distortion of information etc.). For example, one of the important feature of such failure is the activity of the bureaucracy used their possibilities to manipulate supporting resources. Progressive adaptive and intelligent mechanisms for the international regimes functioning are intended for eliminate this activity. Methodology of creating such mechanisms is described by Tsyganov (1990) and Grishutkin& Tsyganov (2001).

One of the important ways to diminish a gap between rich and poor countries is to minimize capital and brain drain. Under consideration is the problem of capital drain. The concept of adaptive control mechanism of international cooperation derived by Tsyganov (1990) is applied to manage capital and brain flows between leading and developing countries.

Adjustment of the international regimes to prevent capital and brain flight may be considered as an indirect way of support to avoid instability. The other way is a straight support of poor country with the aid of international organization such as International Monetary Fund etc. Concept for designing of the adaptive control mechanisms for international cooperation (including rates, incentives, taxes, norms, etc.) had been considered by Tsyganov (1990).

Real international regimes are much more comprehensive and need intelligent control systems. Drawing from experience gained in implementing intelligent control to a varied range of large scale systems, Grishutkin and Tsyganov (2001) highlights the need for a multilevel self-learning and self-organized systems. Particular attention is directed toward adaptations of the widely used self-learning algorithms in an attempt to increase the effective applicability, range of self-organizing control with the aid of artificial intelligence methodology.

The approach suggested to the solution of the problem of adaptive control for the international regimes implemented by Grishutkin and Tsyganov (2001) to global management of the intangible technologies.

4. CONCLUSION

Technological changes provide the technical progress. But most of the countries in the world have no possibility for self-adaptation to these technological changes. Now the globalization processes are under control of world finance&information oligarchy. Its policy provides rupture between rich and poor countries. That may provide global totalitarian net. Mastering capital&information, and control of its flows turn into main instruments of total control and world supremacy. Globalization processes advance on traditional world often in the form of aggression. Societies, political institutions, leaders and citizens most of countries are worry of perspectives of the possibility to lose the independence. Many countries must take a measure to defend their cultures, traditions and values from the alien information influence, and from the evident expansion mechanisms.

REFERENCES

Chestnut, H. and P. Kopacek (1989). Supplemental Ways for Improving International Stability. In: *Report on the IFAC/EPCOM Working Group (WG 7.2) "Control Engineering and International Conflict Resolution"*. Vienna.

Tsyganov V.V. (1990). Modelling of Adaptive Control Mechanism for International Cooperation. In: *Report on the Research Workshop "Models and concepts of interdependence between nations". "Soviet-American Dialogue in the Social Sciences"*, p. 74. National Academy Press, Washington, D.C.

Grishutkin, A.N. and Tsyganov, V.V. (2001). Progressive Adaptive Mechanisms of Globalization. In: *Proceedings of the International Conference on Systems Cognition*. Vol. 2, pp.101-107. IPU RAN, Moscow.

Kononov D.A., Kulba V.V., Shubin A.N. (2001). Stability of socio-economic systems: Scenario investigation methodology. In: *Proceedings of the 8th IFAC conference on Social Stability "The Challenge of Technology Development"*. Vienna.

Kulba V.V., Malugin V.D., Shubin A.N. (2001). Techniques of planning the sets of measures for preventing and overcoming reasons and consequences of emergency situations. In: *Proceedings of the 8th IFAC conference on Social Stability "The Challenge of Technology Development"*. Vienna.

Kulba V.V., Malugin V.D., Shubin A.N. (2002). Globalization – information control processes. In: *Proceedings of the 4th International Conference on Complex Systems*, pp.21-25. Samara, Russia,

ELSEVIER

IFAC

PUBLICATIONS
www.elsevier.com/locate/ifac

ADAPTIVE MECHANISMS FOR IMPROVING INTERNATIONAL STABILITY AT THE TECHNOLOGICAL CHANGES

A.N. Shubin, V.V. Kulba, V.V. Tsyganov

Institute of Control Sciences Russian Academy of Sciences
Profsoyuznaja Str., 65, Moscow, 117342, RUSSIA
E-mail: shoubine@ipu.rssi.ru, kulba@ipu.rssi.ru, bbc@ipu.rssi.ru

T.I. Ovchinnikova

Moscow State Institute of Steel&Alloys,
Leninsky prospect, 4, Moscow, 117936, RUSSIA,
E-mail: tatianka_21@mtu-net.ru

Abstract: Technological changes provide the progress, rises gap between rich and poor countries. Big part of third world has no possibility for adaptation to these changes and need the support. Mechanism of such support should be adaptive, corresponding to the level of this gap in a way used in adaptive control. But "active" bureaucracy may predict the results of the adaptive procedures and to reach own aims. In many cases support given to poor countries had no effect because of the failure of the mechanism used (corruption etc.). Methodology of creating progressive adaptive and intelligent mechanisms for the international regimes functioning is described. *Copyright ©2003 IFAC*

Keywords: active element, adaptive, control, intelligent, international stability, planning.

1. INTRODUCTION

Technological changes provide the progress. But these changes produce also gap between rich and poor countries. Many poor countries of the third world have no possibility for learning and adaptation to these changes and need the support. Developed countries provide some efforts to diminish this gap and to improve international stability. Chestnut and Kopacek (1989) indicate that to reduce the risks and likelihood of the international instability it is possible to apply adaptive control principles for international conflict resolution. Adequate adaptive procedures provide identification of controlled object structure and parameters of environment and, finally, generating the controlling actions making use of current information, obtained from the elements, in order to achieve the optimal state of the system as a whole. The third world support should correspond to the level of this gap between rich and poor countries in a way used in adaptive control systems. From the other side, systemic investigation of the problems dealing with the design of the adaptive systems to control organizations, including as the main component analysis and the account of the human factor effect. It should be understood as an activity manifestation of the people or collectives (elements of control systems) is caused by the availability of their own aims, not necessary coinciding with the goal of the system in its entirely (Burkov and Tsyganov, 1986). Such elements may utilize available information channels connected with the system control center in order to improve the current or future state. From the other side, "active" people deals with the process of this support may predict the results of the adaptive control procedures and to use that knowledge to reach their own aims. In fact, in many cases support given from developed to poor countries had no effect because of the failure of the mechanism used (corruption, distortion of information etc.). For example, one of the important feature of such failure is the activity of the bureaucracy used their possibilities to manipulate supporting resources. Progressive adaptive and intelligent mechanisms for the international regimes functioning are intended for eliminate this activity. Methodology of creating such mechanisms is described below.

2. THE TECHNOLOGICAL CHANGES AND THE INSTABILITY

The technological changes are created by high technologies (high-tech). Advanced technology produces both positive and negative effects of social stability. First of all, the technically advanced systems that are in existence, as well as those being built, can be used effectively to help people to realize the benefits that are possible with the present-day high technology. From the other side, potential of high-tech now is accumulated in a few major leading countries. Multinational corporations from these countries supply high-tech all over the world. To provide new high-tech it is necessary to realize R&D, know-how etc. That needs appropriate intellectual and financial resources. The prices of high-tech goods and services are mainly not competitive prices because of the monopoly on the results of R&D and know-how. From the other side, price of traditional goods, provided by developing countries, are under strong competitions. For this reason investigations and investments in high-tech are in most of cases preferable. Financial and intellectual resources drain from developing to leading countries. The result of the capital and brain drain is the lack of financial and intellectual potentials in developing countries. It becomes more and more difficult to provide R&D, know-how and new high-tech. Therefore technological development produces economical and social rupture between leading and developing countries. That provides likelihood of social crisis and instability, and as a consequence tensions between different countries. Corresponding to the well-known business magazine "Euromoney", edited in UK, in 1999 almost one half of 180 countries in the world (85 countries in March and 87 in September) had credit rating, equal to zero in the same periods. More the one half of all countries (92 from 180) had no access to international bank finance. For example, up to year 2015 total capital flow from Russia is estimated as $600 billions. But Russian budget at the same period is estimated only as $500 billions. This provides obstacles to the realization of a more peaceful and stable set of domestic and international relations.

3. THE CAPITAL DRAIN AND THE INTERNATIONAL REGIMES

One of the important ways to diminish a gap between rich and poor countries is to minimize capital and brain drain. Under consideration is the problem of capital drain. The concept of adaptive control mechanism of international cooperation derived by Tsyganov (1990a) is applied to manage capital flows between leading and developing countries. The mathematical approach used the modified model of the developing active element described by Burkov and Tsyganov (1986). Under consideration is the owner wich provide two business processes take place in two countries. Potential of each business rises in accordance with the equation:

$$q_{it+1} = C_i q_{it} + B_i u_{it}, \quad q_{i0} = q^*_i$$

where q_{it} —potential of i-th business, i=1,2, u_{it} — investments in i-th business in period t, C_i and B_i - positive coefficients, t - number of period, t = 0,1, . The profit of each business is:

$$z_{it} = A_i q_{it}, \quad A_i > 0$$

The total profit of the owner is:

$$z_t = z_{1t} + z_{2t} = A_1 q_{1t} + A_2 q_{2t}$$

Total investments of the owner u_t are equal to the total profit:

$$u_t = u_{1t} + u_{2t} = z_t$$

The purpose of allocation of these investments is to maximize present value:

$$W = \sum_{i=1}^{2} \sum_{t=0}^{T-1} \rho^t u_{it} \xrightarrow[u_t]{} max$$

where ρ is a discount rate, $\rho < 1$, T - horizont for business planning (farseeing) of the owner.

Statement. The country 1 is called as the capital attractor if all the investments are allocated in the business takes place in this country:

$$u_{2t} = 0, \quad t = 1, ,T-1$$

Theorem. The country 1 is the capital attractor if and only if

$$I_{1t} \geq I_{2t}, \quad I_{it} = A_i B_i C_i^t, \quad i = \overline{1,2}, \quad t = \overline{1,T-1}.$$

This theorem gives necessary and sufficient conditions about of the definite direction of the capital flow between two countries. With the aid of this theorem measurement means and criteria for monitoring can be found. In this approach

$I_i = (I_{i1}, ..., I_{iT} - 1)$ is the vectors indicator of the investment climate. If $I_1 \geq I_2$ then it should be income of the investments in country 1. Let us consider the owner operating in different branches of economy. Then in branches in country 1, were investment climate is better, income of investments takes place, etc. Another important indicator (Q_i) is the total quantity of capital drain from the country should be calculate as a difference between outcome and income of the investments in different branches of economy. These two indicators are the functions of the parameters both the domestic and the international regimes such as prices, efficiency of investment, taxes, trade and customs rules etc. Information about I_i and Q_i may provide indication of leading events for various nations that may describe conditions of normal, alert, and emergency operation for various potential trouble spots in the world. Possible courses should be developed for action to resolve situations corresponding to alert and emergency operation. Various possible alternative actions can be explored. Through the use of models, simulations, discussions with knowledgeable experts from both the countries involved, as well as with third-party experts, it should be possible to get various impressions of what may be the possible outcome of alternative actions. In the adaptive mechanism there are special procedures used for adjusting of parameters both international and domestic regimes to control capital flow. For example, simulation of progressive adaptive mechanisms of multistage negotiation based on the new information technologies considered by Tsyganov (1990b). Incentives and motivations for cooperation may be developed. In each of the countries involved the people responsible for the decision-making that causes a capital flow can be provided with various incentives for keeping the capital from the flight. In real situations capital and intellectual flows are managed by the special international procedures and mechanisms. They are called by Young (1982) and Krasner (1983) as an international regimes. Coates and Seamen (1989) indicate that international regimes are closely linked with domestic political and economical procedures and mechanisms. Total mechanism to control capital and intellectual flows includes both international and domestic regimes. For example, detailed description both these regimes, dealing with capital flows through joint ventures, had been given by Tsyganov (1991b).

4. ADAPTIVE MECHANISMS FOR IMPROVING INTERNATIONAL STABILITY

Adjustment of the international regimes to prevent capital flight discussed in the previous item may be considered as an indirect way of support to avoid instability. The other way is a straight support of poor country with the aid of international

organization such as International Monetary Fund etc. Concept for designing of the adaptive control mechanisms for international cooperation (including rates, incentives, taxes, norms, etc.) had been considered by Tsyganov (1990a). This approach is based on the analysis of the problem dealing with designing the adaptive mechanism of functioning (AMF) of the two-level organization included the Center on the upper level and the Agent as the farseeing active element on lower level. The role of Center is played by international organization. The role of Agent is played by the government of the supported country. The AMF includes both adaptive procedure **A** for parameter estimation and procedures of decision making: planning **P**, control **C** and stimulation **S** (see fig. 1). A method to avoid indesirable distortion of information consists of designing the so-called progressive AMF wherein the present value of the Agent long-term goal function corresponding to the solution of the game with the Center increase with the growth of the efficiency of the Agent functioning (Tsyganov, 1986).

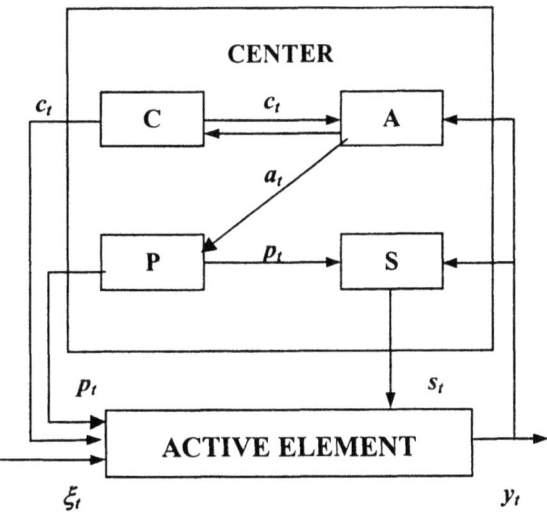

Fig. 1. AMF structure. Here a_t – adaptive parameter, p_t – plan, c_t – control, s_t – stimuli, ξ_t – noise, y_t – output, t – time period.

The designing of AMF caused the development of the approach based on the obtained problem solutions of progressive AMF synthesis. Burkov and Tsyganov (1987) made detail consideration of two AMF main types. The first one is intended for maintaining the processes of the state forecasting, planning and control of the Agent. Under consideration are the adaptive procedures of time series forecasting and designing of regressive model. The second type i.e. rank AMF, is designed to provide information for learning of decision making (classification and pattern recognition). They are used mainly for adaptive estimation of the parameters of the decision making procedure, control and stimulation of the Agent. In some cases both expert adaptive mechanisms derived by Tsyganov (1991b) and intelligent functioning mechanisms considered by Tsyganov and Shishkin (2001) can be used.

4. INTELLIGENT MECHANISMS OF THE INTERNATIONAL REGIMES FUNCTIONING

In two previous items models of indirect and direct mechanisms for diminishing rupture between developed and poor countries had been discussed. Real international regimes are much more comprehensive and need intelligent control systems. Drawing from experience gained in implementing intelligent control to a varied range of large scale systems, Burkov and Tsyganov (1987) highlights the need for a multilevel self-learning and self-organized systems. Particular attention is directed toward adaptations of the widely used self-learning algorithms in an attempt to increase the effective applicability, range of self-organizing control with the aid of artificial intelligence methodology. On the other hand, as a rule the possibility to control of a complex organization in dynamics with no complete information, is based on intelligent information systems (IIS). They realizes model, identification of control objects structure and outside parameters, predictions, and forms the base of control actions on the basis of current information, received from the elements, to attain the systems aim on the whole. To avoid information distortion, passed by the elements to the center of the system, it is necessary to consider the problem of information system designing in a total problem of procedures synthesis such as planning, regulation and stimulation accepted in this control mechanism. In recent years the newly appeared direction in the theory rapidly develops as well as the practice control for hierarchic organizations with a stochastic structure. It has been connected with the designing of the intelligent functioning mechanisms (IFM). The IFM includes IIS, and procedures of planning, regulating and stimulating. In the IFM information received in the process of element functioning is used by the center for decision-making and achievement of the systems aim. These IFM ensure possibility to identify the internal structure of elements and their parameters as well as the utilization of internal elements resources in accordance with the center goal. Tsyganov and Shishkin (2001) indicate that the main types of IFM are:
- Learning functioning mechanisms,
- Self-organizing mechanisms,
- Expert intelligent mechanisms.

Learning functioning mechanisms (LFM) provide the possibility of estimating the parameters of the organizational potential in its dynamics, supplying more information to plan the organization output indices at the account of learning processes. Self-organizing mechanisms should combine learning and planning for output organization indices (the way it is done in LFM) with the control of organization inputs, i.e. direct influence on the potential of an organization. In expert intelligent mechanisms (EIM) the knowledge base is the part of IIS. EIM combines learning with indistinct and qualitative commands from the center and control on these commands basis. To design such mechanisms it is necessary to create hierarchic computerized systems with such intelligent possibility as multi-level learning. The knowledge base consists of:
- main knowledge base includes well-known dependences, any accurate data and the results of individual and collective expertise,
- system of knowledge acquisition functioning in an interactive mode with the decision makers who are responsible for a problem solution and answering the question: "what may actually happened, if ?".

With respect to EIM synthesis the idea of theoretical results consists in the fact that with the sufficiently flexible information usage to solve problems of planning, control and incentives the optimum will be reached. As a result – there appears the possibility of potential identifying on extra basis of received information and gradual slow output to the required level of development. The developed approach is directed to creation of EIM, including procedures of analysis and forecasting of economic elements potential with a high degree of approximation and procedure of decision-making. The approach suggested to the solution of the problem of adaptive control for the international regimes implemented by Grishutkin and Tsyganov (2001) to global management of the intangible technologies.

5. CONCLUSION

Technological changes provide the progress and simultaneously produce rupture between rich and poor countries. Many poor countries of the third world have no possibility for adaptation and need the support. First of all, mechanism of such support should be adaptive. Usually provision of ability to control complex international organization in its dynamics with incomplete information is, as a rule, based on application of adaptive control. From the other hand, control of hierarchic international systems implies the concideration of a special type of human factor - elements activity connected with the availability of the elements own goals. The center obtains information from the active elements in the course of their functioning and uses it for estimating their states to reach the aim of the control. But the farseeing active element may predict the center controlling action and chooses its states in such a way that its effects on the results of state estimation and adaptive control to maximize its own goal function. For this reason the problem of the designing of progressive adaptive functioning mechanism (including estimation, planning, control and stimulating procedures) for the hierarchical international regimes should be taken into account.

REFERENCES

Burkov, V.N. and V.V. Tsyganov (1986). Stochastic Mechanisms of the Active System Functioning. In: *Preprints of the 2ⁿᵈ IFAC Symposium on Stochastic Control.* vol.1, pp.259-263. Nauka, Moscow.

Burkov, V.N. and V.V. Tsyganov (1987). Adaptive information system to control organization activity. In: *Preprints of IFIP TC 8 Conference on Governmental and Municipal Information Systems.* pp.85-95. Budapest.

Chestnut, H. and P. Kopacek (1989). Supplemental Ways for Improving International Stability. In: *Report on the IFAC/EPCOM Working Group (WG 7.2) "Control Engineering and International Conflict Resolution".* Vienna.

Coates, J.F. and S. Seamen (1989). The Relation of Management to Control Technology – Futher Applied Studies in Creative Management of Potential Conflict at International Levels. In: *Preprints of the IFAC SWIIS Workshop "International Conflict Resolution Using Systems Engineering".* pp.47-54. Budapest.

Grishutkin, A.N. and V.V. Tsyganov (2001). Progressive Adaptive Mechanisms of Globalization. In: *Proceedings of the International Conference on Systems Cognition.* Vol. 2, pp.101-107. IPU RAN, Moscow.

Krasner S. (1983). *International Regimes.* Cornell University Press, N.-Y.

Tsyganov V.V. (1986). Adaptive control of Hierarchical Socio-Economic Systems. In: *Preprints of the 4ᵗʰ IFAC/IFORS Symposium @Large Scale Systems. Theory and Applications.* Vol. 2, *pp. 694-698.* Pergamon Press, Zurich.

Tsyganov V.V. (1990a). Modelling of Adaptive Control Mechanism for International Cooperation. In: *Report on the Research Workshop "Models and concepts of interdependence between nations". "Soviet-American Dialogue in the Social Sciences",*p. 74. National Academy Press,Washington, D.C.

Tsyganov, V.V. (1990b). Simulation of Progressive Adaptive Mechanisms of Multistage Negotiations and the New Information Technologies. In: *Preprints of the 11ᵗʰ IFAC World Congress.* Vol.1, pp. 202-206. Tallinn.

Tsyganov V.V. (1991a*). Joint venturing in the USSR.* Radianska osvita, Kiev.

Tsyganov, V.V. (1991b) Expert adaptive mechanisms. In: *Proceedings of the 8ᵗʰ Conference on Systems Engineering.* Vol.1, pp. 92-95. Coventry.

Tsyganov, V. and G. Shishkin (2001). Intelligent Mechanism of the Plant Functioning. In: *Proceedings of the 6-th International Conference CADSM.* pp.119-120. Lvov.

Young O.R (1982). Regime Dynamics: The Rise and Fall of International Regimes. *International Organizations,* **36**, pp.43-52.

ELSEVIER

IFAC
PUBLICATIONS
www.elsevier.com/locate/ifac

THE KNOWLEDGE ENVIRONMENT: BUILDING BRIDGES BETWEEN THEORY AND PRACTICE

Mico Jancev * and Ted O'Keeffe**

Smile IT Solutions, Austria,
E-mail: mjancev@hotmail.com

** *Waterford Institute of Technology, Dept. of Management & Organisation,*
Republic of Ireland,
E-mail: tokeeffe@wit.ie

Abstract: The emerging economies of all Post-Socialist countries continue to express the need for greater interaction with their more developed neighbours. Such relationships are considered crucial for economic survival initially and long-term development as well. On the other hand the developing countries have provided, and still provide, huge amounts of support in different forms to the post-socialist countries. However, casual review and deeper analysis reveal that a significant proportion of these resources do not always deliver on their initial promise for a variety of reasons. This paper endeavours to identify the "missing links" and suggest possible solutions that will go some way to alleviate the concerns identified. *Copyright © 2003 IFAC*

Key Words: Change Management, Knowledge Management, Project Team Building, Problem Solving and Organisational Learning

1. INTRODUCTION

"The only policy likely to succed is to try to make the future... To try to make the future is highly risky. It is less risky, however, than not to try to make it." Drucker (1999)

Constancy of change is the global imperative for organizations in the new millennium. Equanimity is not an option in an environment where transformational change is increasingly the norm. If one takes time out to examine the tremendous speed at which our way of life has changed over the last ten years one begin's to comprehend to enormity of the task ahead. The changes are process, not events; which makes them more difficult to manage. Dynamic change processes require the creation of a special social environment, which cann't be mandated. Initiating such an environment means creating internal awareness, acceptance of the changes and the development of the necessary skills for the change agents assigned the task of delivering the change inititives. This is not an easy task, besides the hard variables such as technology; systems and structures there are the softer issues such as relations, attitudes and organisational culture that need to be addressed as well O'Keeffe (2002:a).

Innovation has an important influence on both the economy and society. For highly developed economies the challenge for those at the top is how to maintain their prolific position and stay ahead of the competition. Successful companies understand that in order to ensure long-term prosperity they must embrace transformational change continuously O'Keeffe & Harrington (2001). While in under-developed countries the learning paradox centres on how to survive long enough to achieve competitive advantage in underperforming economies with poor living standards. The diversity and nature of learning and development issues are very different from one situation to another. For this reason the first step in the change process is an accurate determination of where the organization is on its own unique learning curve and the organizations position relative to internal and external competitors.

It is also clear that real progress will not be possible without unlearning outdated policies and boldly confronting the new reality. Dynamic change in unstable environments is one such issue and though it can generate great uncertainty it should be actively encouraged. What is becoming ever more clear is that the critical change imperatives are neither technological nor economic but are part of a new philosophical approach in the way we think, act and create harmony in our society as a whole. Even in developed economies there is still great potential for improving the interface between advanced technology and human resource implementation O'Keeffe (2001:a).

However, while natural sciences deal with the behaviour of *objects* the social discipline deals with the behaviour of *people* within established frameworks O'Keeffe (2002) the "social universe" has no solid rules of engagement similar to the "natural laws". On the contrary, the assumptions and paradigms from social science are subject to continuous change and those imperatives that were valid yesterday can become invalid overnight or even totally misleading by the following day Drucker (1999).

This paper examines transformational change; knowledge diffusion and technology transfer initiatives from a range of perspectives. In societies where divergent viewpoints exist a programme for prosperity particularly in developing countries will require the integration of a host of divergent interests from all sectors of society. Such a process requires ongoing professional guidance to establish the necessary criteria that will provide the necessary building blocks towards the new vision.

2. POST-SOCIALIST COUNTRIES – CHANGE ISSUES

In parallel with the increasing rate of technology change, the post-socialist countries in Central and Eastern Europe are still trying to come to terms with the great political upheavals that have occurred towards the close of the second millennium. Currently, the informed insider view recognises that the processes of transition, reorganisation and reforming in said countries is more complicated than it initially appears to those looking in from the outside. A series of parallel change processes initiatives such as entrepreneurship, leadership, training and development programmes and ethical issues must be addressed in order that the necessary transformation that will enable these countries to compete on the open market might occur. These learning processes to be effective must be multidimensional, as they require an interdisciplinary approach in their execution (Jancev, 1994; Jancev and Cernetic, 2000; O'Keeffe, 2003). The collapse of the "socialist" system is a natural consequence of ignoring three key factors for successful economic development:

1. Intellectual Capital (knowledge)
2. Laws of Economics
3. Marketing Principles (sine-qua-non for survival).

Real progress in the developing countries of Eastern Europe is being hampered by:

- Each country is still grappling with the thorny issues surrounding economic transition and privatisation of "public" property.
- A considerable number of firms have initiated new technology interventions, yet such activities are been hampered by ineffective leadership more concerned with the process of privatisation rather than performance per se.
- The governments in these countries are still trying to come to terms with the democratic process, which prevents them from adopting a strategic approach to the more important functions of state.
- Recent conflicts have resulted in heavy losses in industrial, public and private property, including infrastructure, which has severely hindered the recovery process.
- Recovery will continue to be slow and piecemeal until such time as the necessary resources urgently required can be acquired cost effectively.
- The industrial firms that previously formed the backbone of these emerging economies have all but collapsed, while the remaining few are desperately trying to adjust to the new reality.
- There exists no coherent policy with regard to development projects funded by outside agencies.

In spite of the many concerns expressed, these emerging countries understand the need to move closer to the developed economies of the European Union in order to stabilise the present volatile situation and more importantly to move forward as an integral part of an united Europe.

At the macro level the European Union is very supportive of faster integration of Eastern European countries into the Union, unfortunately insufficient support is forthcoming to address even the most pressing issues, never mind long-term development. Little or no attention has been given to addressing the *psychological shift* that is required in order to cope effectively with the transformational changes that continue to occur in these emerging economies. However, the evolving dynamic is totally different from that which preceded it in the same way that moving from one political and social system to another is like changing from one energy system organized round a definite source to an alternative organized round knowledge. From this viewpoint it would be absurd to attempt to project the future without analyzing the impact of technological change on the present reality.

The central question has always been: *Who takes care of the common good?* Accepting the notion that there is no universal solution to a particular concern, experience teaches us that some approaches to problem solving are more widely applicable than others. In this context, maybe the Japanes experience could act as a useful template for some of the post-socialist countries. Japanese business post-World War II organized itself around quality circles and empowering staff to take responsibility for decision-making at the lowest level, allowing senior management to pursue strategic business interests. The large Japanese companies learned not to start out with the question: *What is good for business?* But with: *What is good for Japan?* and they asked: *How should business pursue its own interests so that it serves the common good?* Japanes business, throughout the period of that country's reconstruction, identified its ethical responsibility first and foremost and it was this, rather than bureaucratic control was the real "Japanese success story".

In a similar way under performance in post-socialist countries can be linked to under development of the social sciences. Analysts find the only possible and "safe" definition to explain this situation as *"a transitional period"*, where everyone procrastinates until this period has come and gone and it is safe to make "informed" pronouncements Jancev (1999). The reasons why such situations exist are:

- Huge gap between management theory and practice in these countries
- Poor understanding of free market forces
- Lack of team work and interdisciplinary projects
- Insufficient investments in training and development programmes especially at management level.

The tragic consequences of ineffectual leadership structures are still very evident in some areas of the former Yugoslavia and to a lesser degree across the developed world as well. The consequences of ineffectual leadership in post socialist countries are outline below.

- *Ineffective Management.* Too much effort applied to day-to-day activities not sufficient time devoted to addressing strategic issues.
- *Gap between Theory v Practice.* The second most important concern that needs to be highlighted is the knowledge gap between knowledge per se and the effective application of that knowledge.
- *Market Leaders.* Believe that they should have the same rights to market share in a democratic society as they had under the protection of the socialist system.
- *Information Technology.* Computer literacy is a huge concern for the developing Eastern European countries and for the few that have access to information technology are under the misconception that computers can solve all their problems.

Before any real long-term progress can be made in the developing economies of Eastern Europe both political and organizational leaders must come to terms with the following issues:

1. They must accept that to survive thay must embrace continuous change.
2. Social progress is not something that happens it must be actively encouraged.
3. Transition from socialism to a post socialist society will not be easy and must not be imposed as culture, customs, and education standards vary from one location to another.
4. The world is a global village and if organizations or groups wish to become integrated into this society they must pursue a best practice approach in everything they do down to the work of individual employees.
5. Knowledge has no boundaries and become the property of those who are capable of using it.

3. THE KNOWLEDGE ENVIRONMENT

The realities of global competition and increased customer sophistication have focused organizational attention on the need to develop a "learning culture". However, while much has been written on the importance of evolving a learning culture, less attention has been given to understanding in a practical way the characteristics of learning organizations and the ways in which companies can improve their learning systems. The findings suggest that learning organization concepts are perceived as strategically important activities, which can be viewed as a philosophy driven by customer and competitive needs that require executive management commitment and visible support to be successfully implemented O'Keeffe (2001:b)

4. SUPPORTING POST-SOCIALIST COUNTRIES

First world countries have provided, and still provide, huge amounts of support in different forms to the post-socialist countries. However, casual review and deeper analysis reveal that a significant proportion of these resources do not always deliver on their initial promise for a variety of reasons. To understand the reasons for this situation requires a deeper knowledge of the history, political situation, social and economic factors that were the norm in these countries on the one hand while on the other a critical analysis of the support provided is also necessary.

The institutions of post-socialist countries', whether government, business or University continue to directly impact overall performance one way or another. At the same time every one of these institutions have their own concerns that must also be addressed if progress is to be achieved. Experience teaches us that it is one thing to espouse a programme but far more difficult to put such understanding into operational practice in real-life projects. There are three imperatives that seem to surface in turbulent and complex problematic situations:

- All change initiatives are problematic and emerging problems require cost effective solutions;
- Organisations or nations that have not faced up to a given concern will find it more difficult to survive and prosper.
- Leadership in the 21st century will not only have an economical and technical focus but also a social, political and cultural perspective. Effective leaders demonstrate a range of expertise rather than a singular professional qualification.

However, an organisation's success is not solely dependent on good management structures it must have the active support of all stakeholders to eliminate waste and duplication. One example of such waste is the enivatable finding of every study that learning or change initiatives require the active support of a dedicated team in order to be successful. In other words, the creation of an "environment for change" begins with the building a project team composed of professionally competent people capable of addressing interdisciplinary nature and concerns of the issue at hand. This implies that in situations where an organisation or nation lacks the where-with-all to establish such an interdisciplinary problem analysis approach, the chances of finding an acceptable solution will be significantly reduced. This maybe true in cases where the experts members of the team come from different countries, organisations or departments, without sufficient understanding or

empathy for the dynamics at play in any given element of the issue under review, particularly so in situations where some members are meeting others for the first time.

Again contemporary management theory identifies ownership of issues as a key concern and recognises that importance of front line management in problem solving and effective introduction of the new technology and process improvements. While the only valid measure of success or failure of dedicated project teams is the "field test", i.e. successful implementation or resolution of the issue. However, in many cases where international experts are involved the practicalities of implimentation are delegated to the host nation or company executives as they have a finger on the pulse of any given situation.

The third concern underlines the importance of intellectual capital as a critical long-term competitive advantage and as a key-manufacturing resource O'Keeffe (2001:c). This understanding is especially important for former socialist countries that are endeavouring to overcome decades of economic under development. Contemporary management theory recognises that change is pervasive in every sphere of human activity and also discerns the criticality of intellectual capital. From this perspective the question is: *What responsibilities have the various stakeholders for new knowledge acquisition?* It is not sufficient for professionals to apply their knowledge in isolation; that knowledge must be diffused effectively across the organization leading to improved performance or it has no real value. However, in many situations the reality is very different as many so-called experts still espouse the theory of do as I say not as I do. As result there are many instances in which organizations are given outdated advice, wasting precious time and resources and slowing down learning curve effects.

Post-socialist countries, without any doubt, need ongoing support from their more prosporus neighbours and while the nature of this support is not in question the way in which it is provided raises a number of professional and ethical issues.

5. CONCLUSION

Before selecting a development strategy it is necessarily to know the present state of play and the appropriate level of technological development that can be effectively utilised. It is not sufficient to bring people together; far more important is the creation of a working environment where the opinions of diverse groups or individuals (some with hidden agendas) can be channelled effectively towards the creation of a best practice approach to all activities. This dynamic will be difficult to achieve in Post-Socialist countries that have not accepted the need to develop "an environment capably of facilitating transformational change" at all levels. This evolving environment will eventually bridge the gap between theory and practice; embrace effective knowledge diffusion activities (inter-disciplinarily), by linking competence with experience across departments, organisations, regions and countries. This process might be initiated in a specific place (i.e. firm or institution), for an agreed period of time (on a project basis) and with selected individuals (project team). The creation of a social dynamic capable of embracing transformational change in the emerging economies of Eastern Europe could begin with a core group of project leaders, that would develope the necessary training and development programmes systematically for one or more pilot projects that would become the catalyst triggering the ripple effect in performance in the same way that a stone thrown into still water creates wave after wave of movement far beyond the initial impact.

This paper addresses the challenge for both, support providers (developed countries) and support receivers (post-socialist countries) and in doing so raised the following questions:

- Why it is so difficult to achieve dynamic work environments in post-socialist countries?
- Host country concerns working with international project team leaders?
- The role of funding agencies or institutions in resolving these concerns?
- How might individuals be encouraged to take greater responsibility for their actions?

In this new information society the increasing importance of knowledge as a source of competitive advantage is closely linked to the move away from capital per se towards intellectual capital (knowledge) as the key manufacturing resource. The increasing role of intellectual capital in particular and its unique capacity to create knowledge becomes the raison d'être for organisations to develop a capacity for learning O'Keeffe (2003:a). The realities of global competition, the advent of the Internet and increased customer sophistication have focused organisational attention on the need to develop an organizational wide learning culture. Knowledge has no boundaries and can just as easily be diffused across national borders and become the property of those who are capable of using it. This understanding is especially important for former socialist countries that are endeavouring to overcome decades of economic under development.

REFERENCES:

Dragicevic, A. (1990). Decline of Socialism – the End of Mass Society (in Croatian), A. Cesarec, Zagreb

Drucker, P. (1989). The New Realities, Harper&Row, New York

Drucker, P. (1999). Management Challenges for the 21 th Century, Harper&Row, New York

Jancev, M. (1999). No Frontier for Quality, Proceeding of the QMED Conference (Quality Management and Economic Development), September 2-3, Portoroz, Slovenia.

Jancev, M. and Cernetic J. (2000). Change management knowledge can help improving international stability, Preprints of the IFAC SWIIS Workshop on Supplemental Ways for Improving International Stability, Ohrid, Macedonia, May 2000, pp. 45-50.

Cernetic J. and Jancev M. (2000). Implementation of advanced technology in post-socialist countries, Preprints of the 7th IFAC Symposium on Automated Systems Based on Human Skill, Aachen, Germany, June 2000, pp. 251-254.

Jancev, M. (2000). COPIS Approach and Methodology - a key for Business problems solving (in Slovenian), Proceeding of the 9th Annual Quality Assotiation Conference, November 9-10, Portoroz, Slovenia.

Jancev, M. and Cernetic J. (2001). A Social appropriate approach for managing technological change, Preprints of the 8th IFAC SWIIS Conference on Social Stability: The Challenge of Technology Development, Vienna, Austria, September 27-29, 2001, pp. 79-84.

O'Keeffe, T., & D. Harrington, (2001). Learning to Learn: An Examination of Organisational Learning in Selected Irish Multinationals. Journal of European Industrial Training, MCB University Press, **Vol. 25:** Number 2/3/4

O'Keeffe, T., (2001:a). Facilitating Learning within Selected Irish and Foreign Based Multinational Companies. Paper Presented at the Human Resources Global Management Conference at the World Trade Centre: Barcelona, Spain: June

O'Keeffe, T., (2001:b). Learning Characteristics in Multinational Environments: Paper Presentation to the Irish Academy of Management, Magee College, University of Ulster, Derry, Ireland, September.

O'Keeffe, T., (2001:c). Developing Dynamic Multinational Environments. Paper Accepted for Presentation at the British Academy of Management, Cardiff University, Wales, September 7th.

O'Keeffe, T. (2002). Organisational Learning: A New Perspective. Journal of European

Industrial Training: MCB University Press, Vol. 26: Number 2/3/4

O'Keeffe. T., (2002:a) A New Management Development Process: Paper presented to the Strategic Management Initiative Committee of the Irish Government in March 2002 on organisational learning and human resource development in government departments

Keynote Speaker at the Joint EuroSPI & Conquest Conference on Quality Engineering in Software Technology, Theme 2002 was Process Improvement, Methodologies, Technologies, Cultural Factors and Knowledge at the University of Applied Sciences, Nuremberg, Germany 18th - 20th September 2002.

O'Keeffe, T. (2003) Preparing Expatriate Managers of Multinational Organisations for the cultural and learning imperatives of their new environment. *Journal of European Industrial Training*: MCB University Press, **Vol. 27**: Number 5.

O'Keeffe, T. (2003:a) A Human Resource Development Framework capable of locating organisational deficiencies at every level of the organisation. Paper accepted for presentation at the 7th Conference on International Human Resource Management, Limerick 4th - 6th June 2003

www.elsevier.com/locate/ifac

EXPLORING THE DEEP STRUCTURE OF ETHICS IN ENGINEERING TECHNOLOGY DESIGN AND DEPLOYMENT METHODOLOGY

L. Stapleton[1] and M. Hersh[2]

[1] *ISOL Research group, Waterford Institute of Technology*

[2] *University of Glasgow*

Abstract: This paper argues that deep structures are embedded in engineering and technology discourse which work against an inclusive and locally relevant engineering ethics. The paper identifies the need for a new, process-oriented, approach to engineering ethics which enables a dynamic, reconfiguration of ethical issues. This approach must be based upon more locally relevant issues and formally recognise the primacy of the other in relation to the self. It proposes the concept of 'gestalt' as the basis for a theory of engineering ethics. In order to operationalise this theory the paper also submits the Johari Window as a useful device for engineering groups wishing to address local, ethical issues. *Copyright © 2003 IFAC*

Keywords: Engineering theory, ethics, social impact, social stability.

1. INTRODUCTION

It is widely understood by engineers and technologists that power relations are extremely important aspects of any theory of social stability, both at a macro or micro level (Stapleton, 2001; Cernetic & Jerman, 1999; Pederson, 1986; Markus, 1984). Any scientific endeavour embodies the structures and ethos of the society in which it is conceived (Kuhn, 1996). Structuralists show how these structures can be extremely endemic and even subconscious, embodied in the cultural artefacts which surround, and perhaps comprise, scientific progress (Foucault, 1965; Dreyfuss & Rabinow, 1983). By extension decision making about technology development is heavily influenced by the typically western ethos of the surrounding society. Other technologies are ignored and devalued. The affects of underlying power structures are manifested in many ways such as native irrigation techniques in Kenya and minority and women's technologies. Another instance is scientific gatekeeping in which only certain types of science and technology are given official sanction and attempts are made to exclude proponents of 'heretical' ideas from access to resources, including publication in respected

journals. As a consequence, indigenous knowledge, for instance, of edible plants, is disappearing or even suppressed, since it is not recognised as valid or authoritative (Ilkarracan et al, 1995).

Although it is generally assumed that modern western scientific techniques perform better than traditional methods, evidence shows how traditional methods may be better suited to local conditions. For example, traditional techniques of intercropping have been found to give much better yields throughout Africa than the monocropping techniques suggested by 'expert' agronomists (McCorkle, 1989). Again, centuries old small-scale irrigation techniques used by local peoples perform better than irrigation schemes constructed to fit a 'scientific' model (Ikkaracan et al., 1995).

These examples typify what has been described as a colonialist viewpoint in engineering (Bannerjee, 1999). This view combines a lack of respect for the expertise of indigenous people, minorities and women with a lack of respect for the natural environment and remains a central problem in current approaches to the development of technology. Instead of technology being developed in accordance with local needs and expertise and in harmony with the natural environment, current patterns of technology development have resulted in developmental, social and environmental crises. This gives rises to the question of whose interests this pattern of development serves.

2. RELATIONSHIP BETWEEN SOCIETY, TECHNOLOGY AND SCIENCE

There has been considerable discussion of the relationship between society, technology and science, but power relations have rarely been mentioned explicitly in mainstream advanced technology literature. One perspective considers technology to be neutral in itself and its consequences to be determined solely by the nature of particular applications. An almost diametrically opposed perspective, technological determinism (Ellul, 1954; Winner, 1977), considers technology to be all-powerful. In the strongest versions of this perspective technology totally determines the future directions of society in ways that are not possible to resist. Although useful, both these perspectives are too simplistic. In particular they ignore the power relations and dynamics that effect choices about what technology is developed, how it is used and in whose interests they are deployed. These are highly complex processes that are difficult to address according to the positivism underpinning current engineering research (Jervis, 1997).

Discussion of technology has tended to focus on a particular type of development which has taken place largely in the US, Europe and Japan. An important motivation of this type of technological development has been power, often expressed in financial terms, supported by technological determinism, expressed as the belief that a particular development should go ahead simply because it is possible. Achieving positive social change has generally had little influence on this type of development. Recent technological advances are therefore often considered to be linear, rational, western and gendered. This structure ignores other types of technological development that have occurred at different times and places, and in particular developments by indigenous people and women. It also ignores the hidden structures in engineering which have lead, for example, to a gendering of technology (Grundy, 1996; Cockburn et. al. 1993). These structural affects maintain power relations associated with technology including how engineering organisations function (Wilson (2002)). There are also indications that women and men have different approaches to design (GaBe (1983)).

Privileging certain types of knowledge and social behaviour disables individuals who, for whatever reason, including different cognitive processes, find this type of knowledge difficult to assimilate or this type of behaviour difficult to emulate. Approaches to designing technology can be positioned between two poles: design for norms and design for all. Design for norms is based on the implicit assumption that the world's population is white and male, whereas the aim of design is technologies, products and processes that can be used by all sections of society, independent of these factors. The two main models of disability parallel this. In model one the medical model focuses on the individual and the perceived loss of in normal functioning resulting from their disability (Swain et al., 2003), leading to a concern with rehabilitation or trying to make the individual confirm to a particular type of society and infrastructures defined for particular norms. Alternatively, the social model, developed by disabled people in the 1970's and 80's, emphasises the unequal experiences resulting from physical and social barriers (Barnes, 1994), leading to campaigns to change attitudes and remove barriers and recognition of the importance of diversity in society. As well as being based on a political and ideological philosophy that advantages an elite minority at the expense of the majority of the population, design for norms is bad design practice. It leads, for instance, to houses in which no-one want to live or Bhopal and other similar accidents (Hersh et al., 2003;). On the other hand design for all can lead to improvements in quality of life and does not privilege any single social group (Bougie, 1991). Related post-structuralist views emphasise different subjectivities, which consider interpretations to be temporary, specific to a particular discourse and open to challenge (Weedon, 1987). As well as allowing interpretations to be located in a particular time, place, political context and ideology, this type of approach could provide tools to challenge the privilege generally given to dominant ideologies or at least recognise the reasons for this privilege.

Technology design and development are influenced by existing power structures and contribute to developing and further institutionalising particular structures (Baudrillard, 1999; Borgman, 1984). Consequently, technology transfer involves not only the transfer of artefacts and associated 'know-how', but also the unconscious, or deliberate attempts to impose the economic, political and ideological structures in which this technology developed. This can be considered a form of colonisation through technology, which is subtler, but no less insidious than previous attempts (Banerjee, 2001).

3. POWER RELATIONS, TECHNOLOGY AND ETHICS

Consideration of power relations in the development and deployment of technology raises very important ethical issues. These issues are now receiving some attention (Hersh, 2002; Martin et al, 1996; Barbour, 1992) but the literature in this area remains sparse.

Many engineering societies now have codes of ethics or at least codes of professional conduct (Hersh, 1997). However, much technology development theory and practice is still based upon the premise that technology is culturally, politically and socially neutral. Furthermore, it typically ignores the ethical and other responsibilities resulting from the potential power that engineers, and engineering disciplines, have in society.

The consideration of ethical issues frequently focuses on the individual's responsibilities, rather than the development of collective responsibility and organisational and societal cultures of responsibility. While not absolving individuals of ethical responsibilities, a more collective approach, including the development of ethical organisational cultures, would both be more effective and avoid the financial and social penalties paid by individuals who act ethically, for instance by whistle blowing (Hersh, 2001) or refusing to carry out work they consider unethical.

It is apparent that structural affects are subtle, making them difficult for engineers to identify and address. What is needed is a theory of power in engineering that can address these hidden affects at a local level. Also, accompanying practises are required which engineering teams can use to expose these cultural affects in a coherent and open way. This paper introduces the idea of a 'gestalt of power' i.e. a synergistic system of power relations that interact with technology deployment methodologies in deep, but hidden ways. It then proposes a practical technique to address these issues based upon this theoretical approach.

4. GESTALTS OF POWER AND ETHICS

'Gestalt' here refers to a theory of perception developed in opposition to the British 'atomistic' model in which visual patterns were seen to arise from a mosaic of independently existing sensations. The atomistic view represented an attempt to reduce or simplify perceived space into component elements. This is similar to the approach sometimes adopted to engineering ethics. Often, engineering ethics discourse focus on one or two particular issues and attempt to provide a surface level discourse of the issue. For example, Alger, Christensen and Olmsted (1965) represents a traditional approach to ethics in engineering in which ethical issues are addressed as a mosaic of independent issues, such as the ethics of consulting, the ethics of an engineer in industry, government, construction and so on. This approach to ethics details specific sets of surface level issues which western engineers are likely to encounter. It is readily apparent that this is a useful approach as it sets out in some detail guidelines for appropriate professional behaviour.

However, more recently engineers and technologists have attempted to delve deeper, underneath the surface of appropriate behaviours into appropriate attitudes. This is evident in Erman et. al. (1990), the discussion of cultural factors in Martin et. al. (1991) and the discussion of ethical values in Der Vorst (1998). The discussion of these deeper structure issues in engineering has lead to a far richer debate as to what constitutes ethical behaviour. However, to date there remains a theoretical gap in this literature *vis a vis* the organisation of ethical discourse in engineering. At this stage, ethics research needs to find ways in which to organise debate and provide a theoretical framework within which reflective engineers can locate themselves in the grid of complex ethical issues. This needs to be addressed at a personal and inter-personal or, more appropriately, the inter-subjective level.

4.1 Engineering Drawings as a Way of Not Seeing

Debate about technological and engineering ethics often removes the engineer from the context of her invention or his technology. For example, engineering methodologies have built into their very essence this distancing from the locality in which new technology will be implemented. Ihde (1995) argues that the 'visual languages of engineering' (exploded diagrams, drawings etc.) somehow remove the engineer from the context in which the represented objects (technologies etc.) must operate. Therefore, the very approaches used to design and develop new technologies immediately withdraw the engineer from the world in which the new system will be used. It is apparent that the deep structure of engineering visualisations can immediately dis-empower inhabitants of a local context as they disappear from the diagrammatic view. Thus, the ways in which engineers are trained to see (or do not see) the world in which their technologies are deployed has ethical consequences. This also has implications for power-relations in the relationship between engineers and their technologies, and the inhabitants of the social context in which the technologies will be used. This gestalt of power needs to be made explicit and reconfigured in locally appropriate ways. Consequently, devices are needed which can expose the gestalt of ethics and help reconfigure this as appropriate. In this process the gestalt of power will shift through a deeper awareness of my own personal ethical position and its relevance in the local context.

4.2 Towards a Gestalt of Ethics

This paper proposes the theory of gestalt as a means by which we can consider the complex dynamics of engineering ethics. Gestalt, as used by Ihde (1995), implies that the interpretation of an experience changes the experience itself – depictions are interpreted and have meanings, they are not merely objective, engineering diagrams. Gestalt is, therefore, a useful theoretical device for addressing the subjective aspects of ethics i.e. enabling engineering ethics to incorporate a subjective ethics which is culturally-located. By basing itself on a fundamentally post-phenomenological position, this ethics not only emphasises the self (my position in society) it also suggests the other i.e. the need for an inter-subjective approach to ethics. This approach to gestalt emphasises not only what is IN the frame of reference, or what is intentionally perceived (i.e. what is represented) but also what is outside this frame of reference (Schutz (1973)). Thus we can make a shift from 'ego-centric' ethics i.e. ethical discourse centred on 'the engineer', to a focus upon 'the other' – *their* assumptions, thoughts, fears, concerns etc. This enlarges the vocabulary of engineering ethics without diminishing the individuals response to ethical considerations i.e. it avoids ethical discourse becoming so abstract so as to have little meaning on the ground. Through the post-phenomenological approach, and using the idea of a configuration of issues which are personal and inter-subjective, we can argue that there is a gestalt of ethics which any engineer can discover for themselves, whilst simultaneously recognising that there is much which is not perceived but which is also important. Through gestalt theory and post-phenomenology we can move the debate of engineering ethics towards a debate of 'my' ethics and the 'others' ethics/value/life-world and create new shared spaces between the two.

4.3 Primary Dimensions of the Gestalt

It is apparent that primary dimensions of a gestalt of ethics probably include but may not be limited to:
Social identity including ethnic origin, religious persuasion, gender, income, disability status and sexual orientation
social exclusion and decrease of opportunities
environmental issues
granularity of responsibility (e.g. individual, group, societal, institutional and disciplinary)
the distribution of resources and income
intergenerational issues
impacts on development
technological design and deployment issues.
changes to existing power balances
restructuring of time: e.g. availability of employment and leisure opportunities
development and promotion of ethics cultures
It is self-evident that identity factors are central determinants of peoples' expectations and experiences of technology. They also influence available opportunities and the degree of social inclusion, as well as the degree of support encountered in communities of practice. Existing power relations are structurally embedded in identity factors which permit or deny groups and individuals

access to technology. These structural factors lead to technology development approaches that can perpetuate existing power relations and inequalities and injustices. These approaches, and their embedded structures, can be challenged by engineers and others as they develop and re-configure their own ethical gestalts and as they actively contribute to the development of an ethical culture in their communities (De Maria, 1992; Hersh, 2002). These cultural shifts provide a basis for the collective action and solidarity, which is a prerequisite for social change. Individual gestalts are essential, both to provide the basis on which the organisational ethical culture can be built and to encompass differences of experience and perspective.

What is now needed are devices by which engineers and technologists can take the primary dimensions of the gestalt and challenge their own gestalts i.e. their own perceptions of ethical realities. The aim of such devices is to highlight both the ethical problems in a particular context and the ethical gestalt of engineers and technologies, rather than to necessarily obtain 'best practice' solutions. In order to provide a practical basis for this work in the engineering community the next section presents a useful and proven technique from educational research.

6. THE JOHARI WINDOW

The Johari window was originally developed as a diagrammatical device by which people may be made more open to one another and is widely used in reflective learning (Brockbank & McGill (1999)). Figure 1 illustrates the typical Johari window.

The quadrants of the window represent one person in relation to others, with each quadrant revealing awareness of behaviour, emotions and subjective space. Some awareness is shared (inter-subjective) and some is not. Material is allocated to a quadrant on the basis of who knows about it. We will now examine each quadrant in turn.

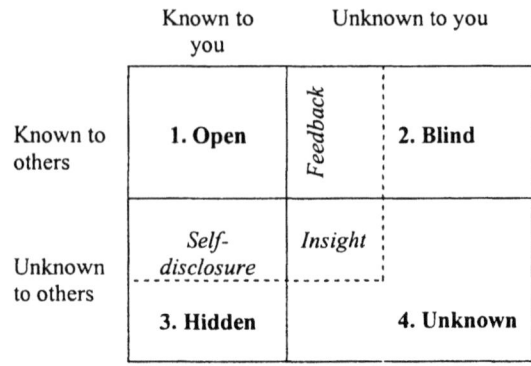

Figure 1. The Johari Window

Quadrant 1: The Open quadrant: behaviour and issues are known to self and others. This is the quadrant that each of us opens to the world and is the basis of most interactions that we willingly display.

Quadrant 2: The Blind quadrant: to that which others see but which I do not. Actions here will be seen in the public gaze – s/he will be aware of some actions (in quadrant 1) and unaware that s/he is displaying other things (quadrant 2). For example, an engineer may not realise that he has inadvertently used a racist expression to another colleague. How the colleague points this out to the engineer and how the engineer reacts will influence how the engineer gets to know about that part of her behaviour of which she was previously unaware.

Quadrant 3: The 'Hidden' quadrant: things I know about myself but which I am unwilling to convey to others. If the engineer discloses issues in this quadrant then they move from here to quadrant 1, reducing the quadrant's 'size'.

Quadrant 4: The Unknown quadrant is something we may get insights into through dreams, psychological counselling and in other ways. This window does contribute to our behaviour but noone, including ourselves, is aware of the deep issues involved.

Now let us review how the Johari window can be used in an engineering work group to raise important ethical issues, maintaining them within their local, intersubjective, context.

6.1 Johari Window in Localised, Ethical Discourse

A group of engineers who wish to explore their own ethical positions can participate in a workshop with an experienced facilitator and use the Johari window to gain potentially deep insights into their own, and others, viewpoints. The workshop will be most effective where major stakeholders and/or a variety of perspectives are brought into the discourse.

The facilitator typically 'breaks the ice' by disclosing something about themselves, thus encouraging others to do the same. It is important that the facilitator ensures that disclosures are appropriate as per Egan (1976) in order to ensure that the ethics, interpersonal psychology and authenticity of the process are protected. The statements that are made need to be authentic as per the following criteria:
Breadth: how much do you want to tell?
Depth: level of intimacy
Duration: amount of time devoted to the process (experience indicates this frequently overruns!)
Target: to whom is information to be disclosed?
Relationships: is it a friend, acquaintance etc.
The situation in which the workshop takes place: for example, private or public place?
There are a variety of guidelines for using this technique and the reader is pointed to Cozby (1973) and Brockbank & McGill (1999) as good sources.

However, amongst the most important are: participants should be encouraged to use statements which begin with 'I' rather than 'you', talk about feelings rather than 'facts', avoid the abstract and remain pertinent and interesting. Self-disclosure can be difficult in western cultural settings where self-disclosure is discouraged amongst, for example, students. Reflecting back is also very powerful in this context. It is important to recognise potential power dynamics between different members of the group, due to identity factors, such as gender or race, different experiences and minority positions. This is in addition to power dynamics resulting from different positions in the organisation and the possibility of discussions that should be confidential to the group being reported back to management. As far as possible a 'safe space' should be created for and the expression personal viewpoints or experiences, and practical barriers to doing so, should be recognised. This device can be accompanied by a semi-structured questionnaire exploring primary dimensions of the gestalt which the group wishes to address (Stapleton (2002)).

The essence of the approach is to expand quadrant 1 in terms of personal ethics through an increased awareness of the engineers' personal values as well as an impression of others' personal positions and ways in which personal ethics impinge upon others.

7. CONCLUSION

It is readily apparent that technology and engineering are not immune from the power relations that impinge upon global society. This raises deep questions about the nature of engineering, what it can achieve, and, indeed, what it can mean to those outside the discipline. The paper recognises a need for theories and techniques that can be co-opted from other disciplines into engineering education and practice. These theories and techniques must recognise that all power is localised and impinge upon individuals. They must also recognise that ethical considerations must be understood in their inter-subjective, localised context. The paper briefly outlines a theory of ethics that is informed by post-phenomenological views of gestalt – the dynamic, re-configurable perceptual framework within which humans perceive (or do not perceive) the world. This theory argues for an ethics based upon context – i.e. both the individual 'I' and the 'other' with whom I live in the world and whom I impact through my engineering. It then proposes one practical approach, the Johari window, as a means for igniting healthy debate by exposing the deep structures underpinning individual ethical positions within the engineering community. This approach makes more explicit individual and group gestalts and recognises the ability for participants to reconfigure this gestalt through their awareness of others.

This approach begins to account for power issues in their relation to engineering ethics. Further work is required to account for when 'I' am the 'other' i.e. for engineers who experience social exclusion either for identity reasons such as being female or black or due to (design) approaches which are not part of the engineering mainstream. This is an ethical research imperative.

REFERENCES

Banerjee, R. (2001). Biodiversity, Biotechnology & Intellectual Property Rights: Unpacking the Violence of 'Sustainable Development', *19th Standing Conference of Organisational Symbolism*, Dublin.

Barbour, I. (1992), *Ethics in an Age of Technology*, The Gifford Lectures Vol. 2, SCM Press.

Baudrillard, J. (1999). *The Consumer Society: Myths and Structure*, Sage: Thousand Oaks.

Borgman, A. (1984). Technology and the Character of Contemporary Life: a Philosophical Inquiry, University of Chicago Press.

Bougie, (1991). The impact of new technologies on the quality of life of persons with disabilities, http://press/coe.int.cp/2001/in361(2001).htm

Cernetic, J. & Jerman, S. (1999). A Soft Change-Management Approach Applied to Information Systems Development, in *Evolution and Challenge in Systems Development*, Zupancic, J., Wojtkowski, W., Wojtkowski, W.G., Wrycza, S., (eds.), Kluwer Academic/Plenum Publishers: NY, pp. 719-726.

Cockburn, C. and S. Ormrod (1993). *Gender and Technology in the Making*, Sage: CA.

De Maria, W. (1992). Queensland whistleblowing, *Australian J. of Social Issues*, **31(1)**, 248-261.

Dreyfuss, H. & Rabinow, P. (1983). *Michel Foucault: Beyond Structuralism and Hermeneutics*, University of Chicago Press: Chicago.

Ellul, J. (1954). *La Technique ou l'Enjeu du Siècle*, Librairie Armand Colin, Paris.

Ezrahi, Y. (1995). Technology and the Illsuion of the Escape from Politics, in Ezrahi,, Mendelsohn & Segal (eds.), *Technology, Pessimism and Post-Modernism*, Univ. of Massachusetts, MA. pp. 29-38.

Grundy, F. (1996). *Women and Computers*, Intellect.

Hersh, M.A. (1997). Environmental ethics for engineers, *Engineering Science and Education Journal*, **9(1)**, 13-19.

Hersh, M.A. (2002). Ethical Analysis Of Automation: A Comparison Of Different Ethical Theories Through Case Studies, *SWIIS '02*, Vienna.

Hersh, M.A. (2002). Whistleblowers – heroes or traitors? Individual and collective responsibility for ethical behaviour, *Annual Reviews in Control*, 26.

Hersh, M.A. and I. Hamburg (2003). *Mathematical Models for Sustainable Development*, Springer Verlag, in press.

Ihde, D. (1998) *Expanding Hermeneutics* Northwestern University Press: Ill.

Ilkkaracan, I. and H. Appleton (1995). *Women's Roles in Technical Innovation*, Intermediate Technology Publication.

Foucault, M. (1965). Madness and Civilisation: A History of Insanity in an Age of Reason, Vintage Books: New York.

Jervis, R. (1997). *System Effects: Complexity in Political and Social Life*, Princeton Univ. Press: NJ.

Kuhn, T. (1996). *The Structure of Scientific Revolutions*, 3rd Ed., Univ. of Chicago: Mich.

Latour, B. (1999). *Pandora's Hope: Essays on the Reality of Science Studies*, Harvard: MA.

McCorkle, C. (1989). Price, preference and practice: farmers' grain disposal decisions in a Burkibane community. In *The Dynamics of Grain Marketing in Burinka Faso*, **vol. ii**, University of Michigan Centre of Research on Economic Development.

Markus, L. (1984). Systems in Organisations: Bugs & Features, Pitman.

Martin, M.W. and R. Schinzinger (1996). *Ethics in Engineering*, edition, McGraw Hill.

Pederson, P. (1986). 'Organisational Power Systems', in Bjorn-Anderson, N., Eason, K. & Robey, D., *Managing Computer Impact: An International Study of Management and Organisations*, Ablex: NJ, pp.127-149.

Stapleton, L. (2001). 'Information Systems Development: An Empirical Study of Irish Manufacturing Companies', Ph. D. Thesis, Department of Business Information Systems, University College, Cork, Republic of Ireland.

Stapleton, L. (2002). Supporting Organisations Learning from Information systems Development: A Research Imperative, International. Symposium on Research Method, RTU, Latvia.

Stapleton, L. (2003). 'Information Systems Development As Folding Together Humans & IT: Towards a revised theory of Information Technology development & deployment in complex social contexts', in Grundspenkis, et. al.(eds.), *Information Systems Development: Advances in Methodologies, Components and Management*, Kluwer: forthcoming.

Weedon, C. (1997). Feminist Practice and Poststructuralist Theory, Blackwell, Oxford.

Wilson, E. (2002). Family Man or Conqueror? Contested Meanings in an Engineering Company', *Culture and Organisation*, pp. 81-100.

Winner L. (1977). Autonomous Technology, Technics-out-of-Control as a Theme in Political Thought, MIT.

ELSEVIER

IFAC
PUBLICATIONS
www.elsevier.com/locate/ifac

BEST PRACTICE IN ORGANISATIONAL LEARNING:
FOUR MULTINATIONAL CASE STUDIES

T. O'Keeffe

ISOL Research Centre, Waterford Institute of Technology
Waterford, Ireland
E-mail: tokeeffe@wit.ie

Abstract The realities of global competition and increased customer sophistication have focused organisational attention on the need to integrate technology within a learning framework. While much has been written on the importance of evolving a learning organisation, less attention has been given to understanding in a practical way the demands advanced technology places on organisational systems, particularly those operating in dynamics environments. There is a growing realisation that long-term organisational success is more dependent on the manner in which it exploits its intellectual capital rather than any state of the art technological interventions. Organisations deploying advanced technology must understand that the interface between man and machine is where most concerns arise and the manner in which such difficulties are managed will influence the real value of advanced technological interventions. *Copyright © 2003 IFAC*

Key Words Learning Organisations, Best Practice, Corporate Culture, Case Study.

1. INTRODUCTION

This paper documents Best Practice in learning organisations as identified by a random sample of executive managers from 126 of the largest multinationals operating in Ireland. These initial findings are supported by four case studies of the top ten multinational companies identified in the earlier study. The case companies were randomly selected from those companies identified as learning organisations from the earlier quantitative research. Indeed, while much has been written on the importance of evolving a learning culture, however, less attention has been given to understanding in a practical way the characteristics of learning organisations and the ways in which companies can improve their learning systems. These are issues that are of particular significance in an Irish context given the growth of emphasis on learning in the knowledge economy and the growing importance of multinational companies within Ireland. Recent statistics show that Ireland enjoys 40 per cent of new

electronics investment into the European Union although it has only 1 per cent of the population of the enlarged E.U. More than 40 per cent of the software sold in Europe is produced in this country making Ireland the second largest exporter of software in the world, after America. Ireland's increasing shift towards a knowledge economy is placing addition demands on human resource development as product and service life-cycle shrink and individuals change jobs more frequently. Furthermore, there is a paucity of research on how organisational learning may be accomplished Seibert (1999) and Tung (1987). While there is widespread acknowledgement of the necessity for organisational learning, there is equal concern with regard to how individual and organisational knowledge can be effectively diffused throughout the firm. Furthermore, the research findings will be located within a new learning model reflecting the complexity of learning organisations in bite size chunks for ease of understanding. This in turn is complemented by the author's more critical comparative assessment of organisational learning practices with self-assessments undertaken by senior managers in each case company.

2. CHARACTERISTICS OF ORGANISATIONAL LEARNING

In determining the characteristics of organisational learning it is important to compare and contrast original case data presented here with previously published studies and the findings of the earlier quantitative research. These three sources are used to develop a definitive model through which an organisation may progress as it exploits its own value chain. In order to evaluate each company's approach to organisational learning, it is first necessary to create a set of criteria against which the companies can be assessed. The existing literature utilised here includes, but is not been confined to, the work of (Marquardt & Reynolds, 1994; Senge, 1990; Lipshitz et al., 1996; Tobin, 1993; Watkins & Marsick 1993). Each of these authors, as might be expected, has his or her ideas as to what constitutes a learning organisation. However, as leading practitioners have developed no satisfactory learning organisation framework to date, this paper will present a learning model capable of locating any company irrespective of size accurately on its own unique learning curve.

Figure 1

The justification for using a set of characteristics to describe the learning process is that it offers a way of making sense of the complex constituents involved in learning. It is only by breaking down organisational learning into its constituent parts that companies can come to terms with the concept per se. While it is important to acknowledge that the characteristics are closely interrelated, it is only by defining individual characteristics that we can understand the learning concept in a meaningful way. The adoption of a comprehensive learning process is a challenging exercise for any organisation. The learning characteristics identified are closely related and are in themselves sub-groups of the four primary

controllers of learning namely: Structure, Strategy, Leadership and Culture. Yet these primary facilitators of learning are not sufficient in their own right, they are sustained by many supporting characteristics that came to light as the study progressed. These learning characteristics are outlined in Fig. 1 below.

3. ORGANISATIONAL LEARNING CHARACTERISTICS

The learning characteristics outlined in figure 1 Above form a critical part of this study however, due to page limitation these characteristics cannot be discussed in a meaningful way other than introducing the reader to the individual sub-elements of each characteristic however, they will be explained at the conference presentation.

Innovation: New Product Development, Transformational Learning, Global Expertise, Empathy and Ethics, Partnerships

Perceived Need: Systemic Learning, Emerging Technology, Contingency approach to Learning, Strategic Planning, Benchmarking.

Executive Challenge: Quality Systems, Shared Mental Models, Stability for Change (Unlearning), Management Development, Executive Commitment

Cultural Imperatives: Empowerment, Cross-Functional Teams, Networking, Stress and Responsibility, Reward Systems

Organisational Wide Learning: 360° Feedback, Synergy, Learning styles, Enrichment Programmes, Customer Responsiveness

Learning Organisations: Best practice, Continuous Learning, Knowledge Repositories, Culture of Excellence, Paradox of Learning

4. ORGANISATIONAL LEARNING RATINGS IN THE FOUR CASE COMPANIES

Table 1 summarises the in-depth analysis of the qualitative research and creates what is, in effect, an organisational learning rating for each company. Each company has been given two ratings. The first rating is derived from the case study research interviews and is an objective assessment based on systematic analysis of best practice at each organisation. The second rating is the average of ratings given by a number of senior managers from each company. Each manager was presented with the various characteristics of the learning model and asked to rate his or her organisation on a scale of 1 to 5 where 1 indicates that the company is ideal or close

Table 1 Ratings of the Four Case Companies

	Company A		Company B		Company C		Company D	
	Author rating	Co. rating	Author rating	Co. rating	Author rating	Co. rating	Author rating	Co. rating
Learning Antecedents	2.50	2.25	2.75	2.50	2.75	2.50	2.75	2.50
Innovation	2.50	2.00	2.60	2.25	2.70	2.50	2.80	2.50
Perceived Need	2.50	1.85	2.65	2.25	2.70	2.50	2.80	2.25
Executive Challenge	2.75	2.25	2.75	2.25	3.00	2.25	3.00	2.25
Cultural Imperatives	2.75	2.50	3.00	3.00	3.10	3.00	3.10	2.75
Organ. Wide Learning	2.75	2.25	2.75	2.25	2.90	2.50	2.70	2.00
Learning Organisations	2.50	2.50	2.70	2.50	3.10	2.60	3.00	2.80
Total Average	2.60	2.25	2.75	2.43	2.89	2.58	2.88	2.44
N = No. of managers in each company	6		6		6		6	

to ideal for learning and where 5 indicates it is unsuitable for organisational learning. This was in all cases, carried out with the assurance of anonymity to encourage respondents to be more open and honest. The ratings of the various managers from the four case study organisations are outlined in table 1 below

In order to give clarity to the ratings of the four companies outlined above in Table 1 an overview is presented in Table 2, which is effectively a simpler, reordered version of Table 1. It gives each company's overall rating rather than the rating for each characteristic of organisational learning. The companies are also reordered to begin with the company gaining the highest author rating and finish with the company with the lowest author rating.

Each firm is ranked on a scale of 1 to 5 where 1 indicates that the firm is an ideal learning organisation and 5 not on the starting blocks yet.

Looking firstly at the managers' own assessments of their companies, it is clear that the standard deviation and the range of scores are quite low. The most highly rated Company Waterford Crystal scored 2.25, while the lowest rated Company C scored 2.58 as ranked by the individual managers, which equates to a 15% lower rating. The low range between the highest and lowest scores can be explained by the fact that these four companies are among ten learning organisation identified from 126 multinational companies studied. However, the results demonstrate that these selected companies believe themselves to be learning organisations even though their individual learning initiatives are not classified as such. Waterford Crystal is ranked highest by both the management and researcher alike with GlaxoSmithkline the next closest to this ranking.

Table 2 Case Companies by Total Company Ratings

Company	Company Rating	Author Rating
Company A Waterford Crystal	2.25	2.60
Company B GlaxoSmithKline	2.43	2.75
Company C (Approval to publish name pending)	2.55	2.89
Company D Honeywell Aerospace	2.44	2.88
Mean Scores	2.42	2.78

Waterford Crystal seems out of line until one understands the international nature of this company's business particularly in relation to issues of quality and product. Towards the end of the case study interviews the training manager at Waterford Crystal announced they had just won the national award for training excellence and was ranked in the top ten European companies in respect of training initiatives. In assessing the four companies, Waterford Crystal and GlaxoSmithKline conform most closely to the ideal of the learning organisation when measured against the learning characteristics identified. The researcher's score in three of the four cases were remarkably consistent at approximately 15% lower ranking, with one exception GlaxoSmithKline, which obtained an author ranking that was only 12% lower when compared with the overall ranking by the various managers from that company. Honeywell Aerospace obtained the next highest ranking of the remaining two companies, with Company C a little further behind. Overall, the companies do reasonably well, with each one achieving high standards in particular areas

5. A DICHOTOMY OF BEST PRACTICE IN FOUR CASE COMPANIES

What Table 2 shows, however, is that the case companies can be placed into two distinct groups. Waterford Crystal scored significantly higher both in terms of its managers' scores and the author's rating, than Company C and Honeywell Aerospace. Of the two groups, the first, comprising Waterford Crystal see Table 3 below is characterised by its centralised, organised, highly focused nature. The second group of GlazoSmithkline, Company C and Honeywell Aerospace tend to be more entrepreneurial and decentralised, but less focused and co-ordinated. This dichotomy is displayed in Table 3 below, which lists the companies in each group and their identifying characteristics. Clearly, this characterisation is by no means perfect and does not hold true for all organisations when looking at each point of differentiation. In particular, as we will see, GlazoSmithKline has over the time scale of this project, moved closer to Group One. Certainly its performance is markedly superior to the Group Two companies. However, the overall picture outlines a distinctive set of differences justifying the creation of the two groups. The implications of the differences between the two groups will be discussed and an attempt will be made to discover why the company in Group One is ahead of those in Group Two. The results are presented is Table 3 below.

Table 3 Groupings within the Case Companies

Group One	Group Two
Company A	Company B, C and D
Centralised	Decentralised
Focused	Unfocused
Co-ordinated	Entrepreneurial
Dominant in its industry/markets	Usually a major player but often number 2 or 3
Products use related technologies	Some-times unrela ted product Technologies
Major shareholder Irish	Major shareholder American
Record of superior performance	*Relatively ordinary performance

6. STAGES OF DIFFERENTIATION AND INTEGRATION

The first point of differentiation between the two groups of companies is that Waterford Crystal (Company A) has a centralised structure with Head Office inclined to keep a tight reign on operations. In marked contrast GlazoSmithKline, Company C and Honeywell Aerospace all have highly decentralised structures in which a great deal of responsibility for day-to-day operations is passed to divisional and/or subsid iary company managers.

Furthermore, GlaxoSmithKline (Company B) has a very small Head Office, which presides over a group of relatively autonomous subsidiary companies. The chief executive of GlaxoSmithKline sees head office in a mainly monitoring role, he explained this by stating, "nothing is controlled by Corporate Head Office instead everything is controlled from the [subsidiary] company". He also regards GlazoSmithKline's recent acquisitions as separate companies but competing for resources from the same pool. Company C describes itself as operating a very decentralised management structure with subsidiary and associated companies, varying in size from 100-2000 people and having a considerable degree of autonomy. Company C's Director of European Operations stated "As long as we continue to outperform our sister plants and the competition on cost, innovation and effectiveness we have a bright future".

Honeywell Aerospace's (Company D) structure is broadly similar to Company C's with a small Head Office and numerous subsidiary companies organised into divisions. The Human Resources Manager described the company as a combination of centralised and decentralised operating structure. He stated, "we're tightly controlled in certain areas and highly decentralised, highly autonomous in others, i.e. we have strong financial reviews..... On the other hand we're very loose in the sense that those people who are driving a given business get to make the fundamental decisions about the business and the way it ought to go in comparison to their competitors". Given that the main responsibility of Head Office in a decentralised organisation is resource allocation Honeywell Aerospace seems to fit the model of a relatively decentralised organisation, albeit with the usual controls from the centre. These differences in organisational structure may be more a result of the different ways the companies have grown in the past rather than being part of a deliberate plan. Waterford Crystal's growth has been largely by acquisitions, which have been quickly and effectively integrated into the corporate structure. The Group Two companies have also achieved significant organic growth, relative to Group One.

Table 4 Importance of Planning and Control in Multinational Organisations

	Financial Control	Strategic Control	Administrative Control
Size of HQ	small	large	large
Control Mechanisms (a) Budgets	Strong	Moderate	Weak
(b) Planning	None	Moderate	strong
Strategic Responsibility	busin ess unit	divisions	corporate HQ
Inter Unit Dependencies	low	moderate	high
Case Companies	A	C and D	B

Source: "The Functions of the HQ Unit in the Multibusiness Firm", in Rumelt, Schendel and Teece, Fundamental Issues in Strategy: A Research Agenda, (Boston, MA: Harvard Business School Press, p. 339: 1994).

There is no clear consensus regarding whether centralisation or decentralisation is likely to lead to higher performance, but these differences in company structure between the two groups may have significant implications for the way that the companies practice organisational learning. Waterford Crystal may find it easier to integrate new work practices company-wide, but it will be less likely to benefit from some of the advantages of decentralisation such as increased speed of decision-making and less bureaucracy. The lack of decentralisation in Waterford Crystal is partially compensated for by its relatively high levels of empowerment, which offer some of the same advantages.

7. IMPORTANCE OF LEARNING ORGANISATIONS

The learning organisation concept is multi-faceted utilising elements from established process improving methodologies but employing them in a more strategic and results orientated fashion. This finding is in line with Lipshitz et al. (1996) who argued that in any organisation, learning occurs at individual, group and organisational level. Although individuals and teams are the agents through which learning occurs, organisations focus primarily on systems level learning. It happens when organisations synthesise and then institutionalise people's intellectual capital, memories, culture, knowledge systems, routines and core competencies. In its original conceptualisation learning is results oriented and hence strong on prescription but weak on implementation. For this reason one should acknowledge the significant body of academic work critical of organisational learning to gain a more balanced perspective of the concept.

Although theorists and practitioners have recognised the strategic importance of learning as a means of providing a sustainable competitive advantage DeGeus (1988) and Stata (1989) few learning organisational programmes have illustrated the tension between exploration and exploitation that is

at the heart of the learning organisation. There are no time limits with regard to learning and learning organisations O'Keeffe & Harrington (2001). Without learning an organisation's future cannot be guaranteed. Learning is a continuous process and it is the rate of learning that ultimately determines the organisation's survival. In any study such as this there will be as many limitations as contributions. However, an important contribution is the framework or learning model developed; if nothing else it will generate debate and further learning. One limitation is that no rules have been devised for managers to transform their companies into learning organisations. There are no neat ideas for implementation that can be easily applied in a wide range of contexts and settings. The view of organisational learning presented here is less straightforward and more complex than those typically offered in consultancy reports and the popular management literature. However, it is likely to strike a chord with practising managers, purely because it conforms more closely to their own tacit understanding of organisational realities.

Other limitations result from practical and methodological constraints. The focus here is on managerial perspectives, intentionally so because of the top down nature of organisational learning strategies. However, some critics may argue that this leads to a somewhat skewed representation of organisational reality. This need not be a major weakness as long as both author and reader are aware of the perspective offered. Attempts were made to introduce a valuable longitudinal element to this study. Unfortunately such attempts were only partially successful. Ideally, longitudinal insights should have been obtained through prolonged contact with managers in the respective case companies. However, time constraints, on both the researcher and interviewees, meant that this was not always possible.

The author began this process in order to gain greater insights into the learning process and now at the completion of this learning cycle, one must acknowledge Lonergan (1961) got it right when he stated "Thoroughly understand what it is to understand and not only will you understand the broad lines of all there is to be understood, you will possess a fixed base an invariant pattern opening upon all further developments in understanding". As the author's understanding unfolds he remains convinced that the only way to comprehend the complexities of multinational companies and the influence of concepts so elusive as the notion of learning organisations is through rigorous intellectual analysis of contemporary management understanding. "Organisational learning is a precondition of effectiveness in that it facilitates companies or groups to attain and maintain superior performance by giving them the capability to regulate themselves towards zero management" O'Keeffe (2001).

CONCLUSIONS

This research highlighted four important controllers of learning organisations namely: structure, strategy, leadership and culture. In addition to the four controllers seven additional key characteristics of successful learning organisations were identified and the complex inter-relationship between these characteristics and the primary controllers with regard to organisational effectiveness was also acknowledged. Each characteristic is defined in the broadest sense possible to allow greater scope for managers to interpret the various elements in a way that maximise the model's potential to individual organisations or situations. By addressing the key characteristics identified in harmony with an organisation's structure, strategy, leadership and culture it is possible to gain greater insight into individual companies, unique operating systems, knowledge and desire to move towards becoming a learning organisation. Each of the four case companies score well on different learning characteristics. However, follow-up interviews found that many managers were unwilling to give particularly high or low ratings to their organisations in areas where they had direct involvement but such concerns were minimised by taking the individual ranking of six senior managers from different functions in each of the four case companies.

REFERENCES

DeGeus, A. (1988) *Planning as Learning*, Harvard Business Review, 1988, M/April pp. 70-74

Garratt, B. (2000) *The Learning Organisation: Developing Democracy at Work*: HarperCollins, London.

Graham, A. (1996) *The Learning Organisation: Managing knowledge for business success*. The Economist Intelligence Unit: New York.

Lipshitz, R., M. Popper& S. Oz (1996) Building Learning Organizations: The Design and Implementation of Organizational Learning Mechanisms. *Journal of Applied Behavioral Science*, **Vol. 32**: pp. 292-305.

Lonergan, B. (1961) In Con O'Donovan's 'Introducing Bernard Lonergan', Mercier Communications, Cork.

Marquardt, M. J. & A. Reynolds (1994). The Global Learning Organization: Gaining Competitive Advantage through Continuous Learning, Irwin: Burr Ridge IL,.

O'Keeffe, T., (2001). An Examination of Organisational Learning in Selected Multinational Companies in Ireland: Dissertation Submitted for the Degree of Doctor of philosophy to the Higher Authority & Training Awards Council, Dublin.

O'Keeffe, T., & D. Harrington (2001) Learning to Learn: An Examination of Organisational Learning in Selected Irish Multinationals. Journal of European Industrial Training, MCB University Press, Vol. 25: Number 2/3/4

Senge, P.M., (1990) The Fifth Discipline: The Art and Practice of the Learning Organisation. Century Business, New York.

Seibert, K. W., (1999) Reflection-in-Action: American Management Association, Organisational Dynamics, Winter, pp. 54-65.

State, R., (1989). Organizational Learning: The Key to Management Innovation. Sloan Management Review, Vol. 30, Spring: pp. 63-74.

Tobin, D. R., (1993). Re-Educating the Corporation: Foundations for the Learning Organization. Omneo, Essex Junction.

Tung, R. L., '(1987) Expatriate Assignments: Enhancing Success and Minimizing Failure. Academy of management Executive, May, pp. 117-126.

Watkins, K. E., & V.J. Marsick (1993) *Sculpting the Learning Organization: Lessons in the Art and Science of Systemic Change*. Jossey-bass, San Francisco.

AUTHOR INDEX